Fundamentals of Agricultural Extension

(BASED ON NEW ICAR SYLLABUS)

THE AUTHORS

Renu Arya completed her graduation and post graduation in Home Science with specialization in extension education and communication management from C.S.A.U. of Ag. & Tech., Kanpur. Presently she is working as a Subject Matter Specialist (Scientist) at Krishi Vigyan Kendra, Behraich under A.N.D.U. Ag. & Tech., Kumarganj, Ayodhya. She has published many research papers/popular articles in national and international journal of repute. She has authored many books entitled Indira Krishi Pratiyogita Darpan, Indira Agricultural Competition Explorer, Objective in Agricultural Extension, Home Science and Extension education. She is recipient of UGC fellow for doing Ph.D. at PAU, Ludhiana. She is actively involved in extension activities of Home Science at Foot Hills of Himalaya.

Anupriya Arya obtained her B.C.A from C.S.J.M., Kanpur and M.C.A. from Kamla Nehru Institute of Technology, Sultanpur and M. Tech. (Computer sciences), Babasaheb Bhimrao Ambedker University (Central university), Lucknow. Presently she is working as a Assistant Professor (Computer Sciences) from Rajkiya Engineering College, Ambedker nagar UP). She has published several research papers/popular articles in national and International journal of repute and authored several books in the field of computer sciences.

Dr. R.L. Arya obtained B.Sc. (Ag & AH) degree in 1984 from C.S.A.U. Ag. & Tech., Kanpur and M.Sc. & Ph.D. in Agronomy in 1986 and 1989 from Division of Agronomy, I.A.R.I., Pusa, New Delhi, respectively. He has vast experience in the field of agronomical research in fodder and pulses under dryland/rainfed areas and tobacco production. He has to his credit of more than 250 research publications in national and international journals. He is the author of more than twenty five books in different aspects of agriculture and agronomy. Being a recipient of four ICAR national awards for outstanding contributions in different aspects of agronomy. He was Senior scientist in ICAR and worked CRIDA, Hyderabad, IGFRI, Jhansi, IIPR, Kanpur and CTRI, Rajahmundry. Presently he is working as a extension worker (District Consultant) in centrally sponsored scheme of National Food Security Mission at Kanpur Dehat.

Dr. Ram Ashish Yadav completed B.Sc. (Ag) and M.Sc. (Ag) in Agronomy from Banaras Hindu University, Varanasi and Ph.D. (Agronmy) form the Chandra Shekher Azad University of Agriculture and Technology, Kanpur. Presently he is working as a Director (Seed & Farms) and as a Project Invetigator at All India Coordinated Research Project on Weed Control in the Department of Agronomy, College of Agriculture, Chandra Shekher Azad University of Agriculture and Technology, Kanpur. He has long experience in teaching of Under Graduate, Post Graduate and Ph.D. Agronomy students with specialization of weed management. He has published several research papers/popular articles in national and international journal of repute and 4 books entitled Weed management on Field crop in hindi and english, weed management in vegetable crops and Usar men dhan ki kheti. He has also guided 10 students of Post Graduate and five student of Ph.D. Dr. Ram Ashish Yadav has been honoured with several awards from different organizations and has received the **"Best teacher award"** from the university. He is also involved in research and extension activities in the field of Agronomy with specialization of weed management.

Prashant Sachan completed his graduation and post graduation in Agronomy from C.S.A.U. of Ag. & Tech., Kanpur. Presently he is research scholar and Ph.D. student in the department of Agronomy at Chandra Shekher Azad University of Agriculture and Technology, Kanpur. He is having brilliant academic record. He has published many research papers/popular articles/pumphlets/leaflets in national and international journal of repute. He is author of two books in the field of agronomy and awarded as a best student at the time of post graduation.

Fundamentals of Agricultural Extension

(BASED ON NEW ICAR SYLLABUS)

RENU ARYA

M.Sc. (Home Science - Extension Education & Communication Management)
Scientist, KVK, Behraich

ANUPRIYA ARYA

M.C.A. and M.Tech. (Computer science)
Government Engineering College,
Ambedker nagar (UP)

R.L. ARYA

(M.Sc. & Ph.D. in Agronomy, IARI, Pusa, New Delhi)
Ex. Senior Scientist
CTRI, Rajahmundry (A.P.)

RAM ASHISH YADAV

M.Sc. (Ag) & Ph.D. (Agronomy)
Director (Seed & Farms)
C.S.A.U.A&T., Kanpur

PRASHUN SACHAN

M.Sc. (Ag) & Research Scholar (Agronomy)
C.S.A.U.A&T., Kanpur

SCIENTIFIC PUBLISHERS

Published by:

SCIENTIFIC PUBLISHERS (INDIA)
5A, New Pali Road, P.O. Box 91
Jodhpur 342 001, India
E-mail: info@scientificpub.com
Website: www.scientificpubonline.com

ISBN: 978-93-90495-50-4 (PB)
eISBN: 978-93-90495-18-4

Printed in India

PREFACE

Agricultural Extension is the important branch of agriculture which deals with the application of scientific research and new knowledge to agricultural practices through farmer education. Agricultural extension is another name for agricultural advisory services. In practical terms, extension means giving farmers-for us, small holders in developing countries knowledge of agronomic techniques and skills to improve their productivity, food security and livelihoods and promoting agriculture as an engines of pro poor economic growth. Extension provides a critical support services for rural producers confronting agriculture: transformation in the global food and agricultural system, including the rise of super markets and the growing importance of standards labels and food safety; growth is non farm rural employment and agribusiness; constraints imposed by HIV/AID and other health challenges that affects rural livelihood; and the deterioration of the natural resource base and climate change.

Fundamental of Agricultural Extension is not only for extension workers/students but it is for researchers, planners and farmers. This book may be useful to student, teachers, researchers, extension workers and developmental officers for reference and easy answering of many complicated questions. It contains simple, easy extension materials, easy to follow tools on the commonly used extension methods and an outline of how extension may be planned and implemented. More innovative methods must be developed to identify systematically farmer's problems and felt needs, and help to formulate and set agricultural research agenda based on such needs and problems. In short, there is a need to develop and improve the conceptual technical and operational methods and tools in order to strategically plan, efficiently manage and scientifically evaluate a problem solving, demand driven and needs based agricultural extension programs. Fundamental of Agricultural Extension covers 21 chapters viz; agricultural extension, new concept of agricultural extension education, extension methods, teaching learning process, communication, communication processes, audio-visual aids, technology demonstrations, adoption and diffusion process, extension programme planning, information and communication technology, information technology for agricultural development in India, planning and evaluation of extension programmes, agricultural and rural development programme in India, participatory rural appraisal, agricultural extension systems in India, management in agricultural extension organization, organizational structure of extension services and role of women in agriculture are included in brief and simple language . The entire book is prepared in most simple, clear and talking language so that the concept of book could be easily understand by the readers.

This book is also useful for competitive exams like IAS, IFS, PCS, Banking services, states and national levels of different competition and getting of admission of post doctorate and Ph.D. in agricultural subjects. Finally, I have belief and hope that this book will fulfill the requirement of students of agricultural stream.

Author shall be grateful to the various books, magazine, newspapers, newsletters, periodicals, journals etc are rendered help for bringing out of this book. The valuable contributions of different persons who are directly or indirectly involved for preparation of this book are hereby gratefully acknowledged. Authors are highly thankful to M/s Scientific Publishers (India), Jodhpur, Rajasthan and their staff members for printing in the form of book.

Being first edition of book, there could be some shortcomings for which valuable and informative suggestions will be gratefully appreciated, so that these shortcomings may be incorporated in forthcoming edition. The publisher shall also be grateful for bringing to their notice of any suggestions for further improvement of this book.

Authors

CONTENTS

Chapter - 1 **Agricultural Extension**

Agricultural Extension

- Extension (Latin word) = Ex (Out) + tensis (Stretching)
- The word 'Extension' was used first time in USA
- Extension activity was started first time in USA
- First extension work was done in 1785 in Philadelphia state of USA when few farmers got together and organised to sell their produce.
- In 1862, Morril Land Grant Act was passed in USA and signed by the president Lincon. The objective of this act was "All the agricultural institutes would pass on their innovations to rural people".
- In 1885, the concept of demonstration was given by Dr. Seeman A. Knapp. He generated a new variety and demonstrated on one million acre farm in Lausannia.
- Dr. Seeman A. Knapp is known as father of demonstration.
- On the basis of American pattern *i.e.* Land grant college pattern, the first agricultural university at Pantnagar was established in India in 1962.
- The study of extension education as a course first time started at Sabour (Bhagalpur) in Bihar in 1956.
- Extension is an integral part of agriculture and rural development programme in India.
- Extension education is an integral part of teaching in agriculture, veterinary and animal science, dairy, fishery, forestry, sericulture, home science, rural development, social work etc.
- There is scope for introducing extension in other discipline as a subject of teaching, particularly which involve educating and motivating people and assisting them in their profession.
- Extension is an educational, cooperative, collaborative, democratic, persuasive and never ending process that generates harmony, leadership, effective communication, participation and involvement of the people in organizing self help activities and projects.
- It is a system of learning by doing, seeing believes at an individual level and mutual learning at group level.

- Extension as a practice has a long history both ancient and modern. It can be argued that extension dates back to times when human civilization started farming.
- There is archaeological evidence that people were sharing information on better methods of agricultural practices at that time. An example of this is the Mesopotamian civilization, a success story that led to the transformation of agriculture and rural areas in many parts of the world.
- Extension has evolved in different forms with different purposes over time.
- In the end the extension has evolved into a formalized public service that receives huge budget and human resources and is undertaken by governments with the objective of improving the abilities of rural people to adopt technologies and new practices so as to adjust to changing conditions and societal needs.

Definitions of Agricultural Extension

- "Agricultural extension is a bridge that fills the gap between agricultural research stations on the one hand and the farming population on the other by the establishing a suitable teaching organization at various levels of administration".
- Extension work is an out of school system of education in which adult and young people learn by doing. It is partnership between the government and the people, which provides service and education designed to meet the people. Its fundamental objective is the development of the people (Kelsey and Harne, 1963).
- Extension is a programme and a process of helping village people to help themselves, increase their production and to raise their general standard of living (D. Ensminger, 1961)
- It can be defined that extension as the increased dissemination of useful knowledge for improving rural life (H.W. Butt, 1961).
- Extension is a two – way channel; it brings scientific information to the village people and also takes the problems of the village people to the scientific institutions for solution, it is a continuous educational process, in which both learner and teacher contribute and receive (B. Rambhai, 1958).
- Agricultural extension is a professional methods of non –formal education aimed at inducing behavioural changes in the farmers for increasing their income through increased production and productivity by establishing firm linkages with research for solving farmer's problems ensuring adequate and timely supply of inputs and usage proven methods of communication for speeding of the process of diffusion and adoption of innovations (V.C. Sanoria, 1986).

Objectives of Agricultural Extension

The objectives of extension can be expressions of the end towards which our efforts are directed. Even if the fundamental objective of extension is the development of the people, the specific objectives that have been developed over time in the history of extension can be categorized as follows:

Fundamental objective

The fundamental objective of extension is the development of the people or the "Destination man". In other words, it is to develop the rural people economically, socially and culturally by means of education e.g., to increase socio-economic status and standard of living of Indian farming Community.

The fundamental objectives of the extension are:

1. To assist people to discover and analyse of their problems, their felt and unfelt needs.
2. To develop leadership among people and help them in organising groups to solve their problems.
3. To disseminate information based on research and/or practical experience, in such a manner that the people would accept it and put it into actual practice.
4. To informed the farmers problems from time to time to the research workers, so that they may offer solutions based on necessary research.
5. To assist people in mobilising and utilizing the resources which they have and which they need from outside e.g., to increase the production and productivity of paddy in India.

- To disseminate research information of economic and practical importance in a way people would be able to understand and use.
- To collect and transmit feedback information for solving management problems.
- The dissemination of useful knowledge and information relating to agriculture, including the use of improved technologies and improved cultural practices in a variety of farming practices
- To improve all aspects of rural people lives within the framework of the national development policies and people's need for development

Working objectives

It focuses on specific activity of a specific group in a selected geographic area e.g., to increase the yield of Pusa early dwarf of the tomato among the tomato growers of Kalyanpur block in Kanpur District. The major objectives of extension may also be categorized as follows:

i. Material - increase production, income.
ii. Educational - change the outlook of people or develop the individuals.
iii. Social and cultural - development of the community

Scope/areas of agricultural extension

The following nine areas indicate the scope of agricultural extension are:

1. Increasing the efficiency in agricultural production
2. Increasing the efficiency in marketing, distribution and utilization of agricultural inputs and outputs.

3. Conservation, development and use of natural resources.
4. Proper management on the farm and at home.
5. Better family living.
6. Youth development
7. Leadership development
8. Community development.
9. Improving Public affairs for all round development.

The following statements will further amplify the scope of extension. These are:

1. Extension is fundamentally a system of out of school education for adults and youths a like.
2. It is a system where people are motivated through a proper approach to help themselves by applying science in their daily lives, in farming, home making and community living.
3. Extension is education for all village people.
4. Extension is bringing about desirable changes in the knowledge, attitudes and skills of people.
5. Extension is helping people to help themselves.
6. Extension is working with men and women, boys and girls, to answer their felt needs and wants.
7. Extension is teaching people what to want (*i.e.*, converting unfelt needs into felt needs) as well as how to work out ways of satisfying these wants and inspiring them to achieve their desires.
8. Extension is teaching through 'learning by doing' and 'seeing is believing'.
9. Extension is working in harmony with the culture of the people.
10. Extension is a two – way channel; it brings scientific information to village people and it also takes the problems of the village people to the scientific institution for solution.
11. Extension is working together (in groups) to expand the welfare and happiness of the people with their own families, their own villages, their own country and the world.
12. Extension is a living relationship between the village workers and the village people. Respect and trust for each other, sharing of joys and sorrows, results in friendship through which village extension work continues.
13. Extension is development of individual in their day to day living, development of their leaders, their society and their world as a whole.
14. Extension is a continuous educational process in which both learner and teacher contribute and receive improved technologies of various aspects.

Basic philosophy of agricultural extension

Philosophy is the pursuit of wisdom, a body of general principles or laws of a field of knowledge. Philosophy of a particular discipline would furnish the principles or guidelines with which to shape or mould the programmes or activities relating to that discipline.

The philosophy of extension work is based on the importance of an individual in the promotion of progress for rural people and for the nation. Extension Educators should work with people to help them, develop themselves and achieve superior well-being. Sound extension philosophy is always looking ahead.

1. Self help
2. People are the greatest resource
3. Programme of change
4. Cooperative effort
5. Democratic nature means development of people by the people and for the people
6. Two channel of knowledge and experience
7. Programme if based on creating interest by seeing and doing
8. There is voluntarily cooperative participation
9. Persuation and education of people
10. Programme is based on attitude and values of the people with whom it works
11. It is never ending process

According to Kelsey and Hearne (1967) the basic philosophy of extension education is to teach people how to think, not what to think. Extension's specific job is furnishing the inspiration, supplying specific advice the technical help, and counselling to see that the people as individuals, families, groups and communities work together as a unit in "blueprinting" their own problems, charting their own courses, and that they launch forth to achieve their objectives.

Concept of agricultural extension

Extension is a term which is open to a wide variety of interpretations. This understanding will be based on past experience and the particular type of extension service in which the agent is working. In other words, there is no single definition of extension which is universally accepted or which is applicable to all situations. They also all underline extension as an educational process which works with rural people, supports them and prepares them to confront their problems more successfully.

The term extension may be examined by looking at a number of statements that have been written about it.

- Extension is an informal educational process directed toward the rural population. This process offers advice and information to help them to solve their problems. Extension also aims to increase the efficiency of the family, farm, increase production and generally increase the standard of living of the farm family.

- The objective of extension is to change farmers' outlook toward their difficulties. Extension is concerned not just with physical and economic achievements but also with the development of the rural people themselves. Extension agents, therefore, discuss matters with the rural people, help them to gain a clearer insight into their problems and also to decide how to overcome these problems.
- Extension is a process of working with rural people in order to improve their livelihoods. This involves helping farmers to improve the productivity of their agriculture and also developing their abilities to direct their own future development.

Functions of Agricultural Extension

The functions of extension is to bring about desirable changes in behaviour by means of education; changes may bring about in their knowledge, skill, attitude, understanding, goals, action and confidence.

1. **Change in knowledge:** means change in what people know.
2. **Change in skill:** means change in the technique of doing things.
3. **Change in attitude:** it involves change in the feeling or reaction towards certain things.
4. **Change in understanding:** means change in comprehension.
5. **Change in goals:** is the distance in any given direction one is expected to move in a given period of time.
6. **Change in action:** means change in performance or doing thing.
7. **Change in confidence:** it involves change in self reliance.

Principles of agricultural extension: The universal truth in extension which have been observed and found to hold good under varying conditions and circumstances are presented hereunder:

1. Principle of cultural difference.
2. Principle of grass root organization
3. Principle of indigenous knowledge
4. Principle of interests and needs.
5. Principle of learning by doing.
6. Principle of participation.
7. Principle of whole family
8. Principle of leadership
9. Principle of adaptability
10. Principle of satisfaction.
11. Principle of evaluation.

1. Principle of cultural difference

Extension work is based on the cultural background of the people with whom the work is done. This means that the extension worker has to know the level of the knowledge, and the skills of the people, methods and tools used by them, their customs, traditions, beliefs, values, etc. before starting the extension programme.

2. Principle of grass roots organization

The programme should fit with the local conditions. The aim of organizing the local group is to demonstrate the value of the new practices or programme so that more and more people would participate. A group of rural people in local community should sponsor extension work.

3. Principle of indigenous knowledge

People everywhere have indigenous knowledge systems which they have developed through generations of work experience and problem solving in their own specific situations. The indigenous knowledge systems encompass all aspects of life and people consider it essential for their survival.

4. Principle of interest and needs

Extension work must be based on the needs and interests of the people. These needs and interests differ from individual to individual, village to village, block to block and state to state.

5. Principle of learning by doing

Learning remains far from perfect, unless people get involves in actually doing the work. Learning by doing is most effective in changing people's behaviour. This develops confidence as it involves maximum number of sensory organs. People should learn what to do, why to do, how to do and with what result.

6. Principle of participation

Extension helps people to help themselves. Good extension work is directed towards assisting rural families to work out their own problems rather than giving them ready-made solutions. Actual participation and experience of people in these programme creates self-confidence and also they learn more by doing.

7. Principle of whole-family

Extension work will have a better chance of success if the extension workers have a whole-family approach instead of piecemeal approach or separate and unintegrated approach. Extension work is, therefore, for the whole family, *i.e.* for male, female and the youth.

8. Principle of leadership

Extension work is based on the full utilization of local leadership. The selection and training of local leaders to enable them to help to carry out extension work is essential to the success of the programme. People have more faith in local leaders and they should be used to put across a new idea so that it is accepted with the least resistance.

9. Principle of adaptability

People differ from each other, one group differs from another group and conditions also differ from place to place. An extension programme should be flexible, so that necessary changes can be made whenever needed.

10. Principle of satisfaction

The end-product of the effort of extension teaching is the satisfaction that comes to the farmer, his wife or youngsters as the result of solving a problem, meeting a need, acquiring a new skill or some other changes in behaviour. Satisfaction is the key to success in extension work. "A satisfied customer is the best advertisement."

11. Principle of evaluation

Extension is based upon the methods of science, and it needs constant evaluation. The effectiveness of the work is measured in terms of the changes brought about in the knowledge, skill, and attitude and adoption behaviour of the people but not merely in terms of achievement of physical targets.

Agricultural extension principles are essential in promoting an extension programs. Extension principles are paramount for both the extension agents and the communities in which an extension program are to be implemented. For example the principle of culture difference is crucial to the implementation as culture varies from community to community as the tradition, norms and values of communities must be given a serious consideration.

Importance of Agricultural Extension

The importance of agricultural extension are as:

1. Extension education plays major role in bringing desirable change in rural people.
2. The aim of extension education is community development.
3. It is utilized to make scientific methods available to the rural people, so that they can raise their agricultural production and their standard of living.
4. Its main objective is to bring necessary change in the beliefs or views of people.
5. Extension education is primarily aim for the rural development.

Dimensions of extension

There has been an evolution of the concepts over time and they are as follow:

1. Altruistic dimension
2. Educational dimension
3. Communication dimension
4. Behavioural dimension
5. Technology dimension

6. Research dimension
7. Inputs dimension
8. Income dimension
9. Professional dimension
10. Management dimension

Management of extension

The process by which people, technology, job tasks and other resources are combined and coordinated so as to effectively achieve extension objectives for agricultural development.

Types of extension

There is no one universal type of extension but a variety of activities and approaches which can be called extension. The agriculture is the basis of a rural economy, agricultural extension is the most common type of extension to be found in rural areas. But the areas of knowledge and new ideas that farmers and their families require are not restricted to agriculture. There are other aspects of family life in which new knowledge and practices can lead to improvement. Extension is any activity that works with farmers and their families in order to improve the economic and social conditions of their lives and to develop their ability to take responsibility for their own future development.

1. Agricultural extension

An agricultural extension service offers technical advice on agriculture to farmers, and also supplies them with the necessary inputs and services to support their agricultural production. It provides information to farmers and passes to the farmers new ideas developed by agricultural research stations. Agricultural extension programmes cover a broad area including improved crop varieties, better livestock management, improved water management and to control of weeds, pests and diseases. Where appropriate, agricultural extension may also help to build up local farmers' groups and organizations so that they can benefit from extension programmes. Agricultural extension, therefore, provides the indispensable elements that farmers need to improve their agricultural productivity.

2. Non-agricultural extension

In the absence of a collective term to cover the other types of extension, it is convenient to refer to them all as non-agricultural extension. This term includes all activities and efforts not directly related to agriculture or livestock production, but which are important to the farm families. Home economics, family health and nutrition, population education and community development are all non-agricultural extension activities.

3. Rural extension

It covers many aspects of rural life. Home economists and community development workers, therefore, are extension agents who deal with farm families in the same way as agricultural extension agents. In fact, it is becoming increasingly common to talk of rural extension as a collective term which brings together all agricultural and non-agricultural extension activities. The feature common to both types of extension is that they work with families in rural areas and deal with problems in a rural environment. The understanding of extension, the methods used by extension agents, the planning processes involved and the qualities and skills required by agents are factors relevant to all forms of rural extension. The content and subject-matter may be different, but the same general principles apply in both types of extension.

AGRICULTURAL EXTENSION IN INDIA - THE NEXT STEP

The arrangements for agricultural extension in India have grown, over the last five decades, in terms of activities, organisational types and available manpower. Public sector extension, represented mainly by the State Department of Agriculture, continues to be the most important source of information for the majority of farmers. Activities of other extension agencies, be it Non-Governmental Organisations (NGOs), input agencies, mass media, research institutions or farmers associations, though increasing, are still restricted to certain regions, crops and enterprises. The performance of public sector extension is under scrutiny for quite some time and questions

The shifting emphasis of Indian agriculture towards diversification, commercialisation, sustainability and efficiency has made it necessary for the state extension organisations to critically examine their extension approaches. Department of agriculture in several states made changes in some of their approaches towards the late 1980's as the Training and Visit System of Extension was coming to an end.

The changing nature of Indian agriculture shrinking resource base

The land and water resource base for an average farm holding has declined considerably during the last five decades. Due to increasing resource degradation is the inappropriate and unscientific use of land and irrigation water. Degraded lands are either going out of cultivation or are being used for growing low value crops and trees/plantation crops.

Changes in demand and consumption pattern

Per capita food grain consumption for food declined somewhat over the past three decades, while the consumption of fruits, vegetables, meat, fish, eggs and dairy products increased. The demand for livestock products has been increasing rapidly during the last two decades. Increasing per capita income and changing lifestyles are expected to further increase the demand for milk, fruits and vegetables. Rapid growth in livestock demand would push demand of cereals for livestock feed. Assuming a modest growth in per capita income of 2 per cent, the 291.95 mt total food grain production in 2019-20, whereas, the demand in 2025 is projected at 333.0 million tons. The increased demand has to be primarily met through increase in productivity gained through increased application of knowledge by the farmers.

Changing farming systems

Farmers require a different type of support (training, problem-solving consultancy, marketing advice etc.) for growing many of these crops, than simply information on technology. Declining public investments in agriculture like investments in irrigation, rural roads, rural electrification, storage, marketing, agricultural research and education, land development, co-operation etc) in real terms since mid-seventies have been declining consistently in all the states. Emerging trends and regional variations in agricultural investments and their implications for growth and equity. Extension may have to support farmers in this endeavour. The increasing pressure on research funds to find technological solutions to more diverse problems necessitates serious efforts in research prioratisation and targetted technology development. Extension need to assist and direct researchers in problem focussing and evaluating technological options.

International developments

Liberalisation of agricultural trade, consequent to the WTO agreements has resulted in new opportunities and threats to Indian agriculture. India is likely to gain in some crops, but consistent efforts for improving quality (to meet international standards) and increasing cost effectiveness (increasing productivity, achieving cost reduction) in these crops/products are essential to achieve these. Liberalisation of agricultural imports, which would gain further momentum in the months to come, would subject our producers to competition from outside. There is an urgent need to increase the competitiveness of Indian agro-products.

The changing needs of farmers for support from agricultural extension

Due to changing face of agriculture, farmers have to make a number of complex decisions. Most relevant of them are as follows:

a. What technological options could be used profitably in his/her situation keeping in view the potential resource constraints in terms of land, capital, labour and knowledge?
b. How to manage the various technologies e.g. how to make optimal use of new inputs in his farm?
c. How and when to change his farming system e.g. diversifying from crop production to mixed farming or vegetable or animal production
d. For which type of products, is there a good demand in the market?
e. What are the quality specifications he should achieve to get good value for his produce and how to achieve?
f. How, when and where to buy inputs and sell products?
g. How to make decisions collectively on resource use and marketing?
h. How to find quickly the most relevant and reliable knowledge and information?
i. What are the feasible off-farm income generation options available for him and how far he could depend on them?
j. What are going to be the implications for his farming if the input subsidies are phased out and/or if the trade in agriculture is liberalised?

To make good decisions, farmers need information from different sources and often need help to integrate them. Farmers are presently receiving information from extension mainly on technologies generated in research stations and passed on to extension.

Changes in extension approaches

After the close of the World Bank supported National Agricultural Extension Project (NAEP), the central support to the state extension services initiated different extension approaches. Rajasthan adopted the group based approaches to extension and presently the village extension workers operate mainly through Kisan mandate (group of 20 farmers). The state has also been encouraging NGOs to participate in extension activities and has been contracting out some extension activities to them, particularly in the far flung and remote areas where public extension is comparatively weak. Maharashtra adopted the single window system from July 1998, by merging the Departments of Agriculture, Soil and Water Conservation and Horticulture at the operational level. Kerala decentralised the functioning of the Department of Agriculture way back in 1987 by creating offices of DoA in all Panchayats. In 1989, the state initiated the group approach for extension in rice farming and this was subsequently extended to other crops. Punjab had been adopting the SAU-Farmer Direct Contact method over the past two decades and has also upgraded all front-line extensionists to graduate level. Andhra Pradesh Agricultural University has also established District Agricultural Advisory Technology Centres in all the districts for technology refinement, diagnostic visits and for organising field programmes in collaboration with DoA and allied departments.

Different approaches are also being tried in several projects in specific districts. For instance, the Agricultural Technology Management Agency (ATMA) model is presently implemented in 640 districts of 29 states and 3 union territories in the country. This is a bottom-up approach based on Strategic Research and Extension Plans (SREP) prepared at the district level based on Participatory Rural Appraisal/approach. Integration of the activities of agricultural and allied departments and other organisations such as KVKs is expected to be achieved through ATMA at the District level. Similar approach is also tried under Uttar Pradesh Diversified Agricultural Support Project (UPDASP). Apart from these, in several states, various approaches for organising farmers groups as Self Help Groups, are being tried mainly as part of external technical assistance to governments/NGOs. Krishi Vigyan Kendras, continues to be the main source of training for farmers and agro-based entrepreneurs at the district level.

The need for making the system broad based, demand driven and farmer accountable has been widely recognised. The need for linkages between other extension providers, prioritising public interventions, cost-recovery, contracting out services, higher use of mass media and Information Technology etc are also gaining acceptance.

Linkages

Extension has to maintain effective linkages with several systems, only the Research Extension linkages capabilities of the KVK and the quality of linkage have been so far emphasised. However, several measures have been taken to improve this linkage. "Information flow has been mostly top-down" and "the weak feedback has not resulted in any fundamental change in the way research priorities are set at the research stations". Apart from linkages for obtaining technology, the capabilities of the extension agency to assess and refine them for its integration in their knowledge base has been very weak, mainly due to lack of qualified staff. The contribution of KVKs to do this important activity would eventually depend upon the line departments. In many cases, KVKs were found to be wanting in both. Efforts to develop functional linkages with other systems had been adhoc with no real outcome. With its main emphasis on transfer of technology and implementation of schemes which are input distribution related, the extension system never did take the other systems seriously though its performance did depend upon all these factors. As Mosher (1966) wrote more than three decades ago, "agricultural development requires markets for farm products, a constantly changing agricultural technology, local availability of supplies and equipment, production incentives which make it profitable for farmers to produce more and transportation to and from the village".

An extension organisation is a knowledge intensive organisation, which is involved in the production and dissemination of knowledge. Hence the success of this organisation depends to a large extent on knowledge "management. Major role of its managers is to ensure that

- it gets relevant knowledge where ever this is produced,
- the staff members use their creativity to acquire/develop new knowledge,
- all staff members have access to all knowledge which is available in the organisation,
- one learns from experience on how to develop more effective extension methods,
- there is a social climate which stimulates sharing of knowledge and a critical analysis of the knowledge developed or used by colleagues, successful agricultural extension agencies are learning organizations.

The most important challenge for the future extension managers would be the Management of Knowledge. The success of a farmer in the years to come is going to be primarily dependent upon his level of knowledge. The real prices of agricultural products are falling, because knowledge makes it possible to produce products with less land, labour and other resources. In many countries, farmers, who are farming at a knowledge level a good farmer had 10 years ago, have to go out of business because they can no longer compete with more competent farmers.

Reforms

To be effective and to remain relevant in the years to come, the state extension departments should initiate the following structural reforms in the organisation.

- Strengthen its understanding on matters with respect to technology, markets, prices, demand and policies. Departments have to either recruit specialists or have to hire the services of professionals in these areas.
- Recruit better qualified staff - States have to initiate (as Punjab has done) measures to ensure that, the minimum essential qualification for an extension staff should be a graduation in agriculture. In total the states employ some 1,10,000 extension staff of whom presently only around 20 per cent are graduates.
- Improve social science skills of extension personnel. Apart from technical skills, "extension personnel needs several social science skills with respect to need assessment, group formation, negotiation, conflict resolution, mobilisation, management of CPRs, use of IT, data collection, analysis and documentation" .

- Increase the allocation for operational expenditure. "Allocation of operating expenses in State Departments of Agriculture is around 15% whereas a fully functional extension system should have 30-35% of its total expenses as operational expenses".
- Decentralise the operations of the department and provide flexibility to field level officers to decide appropriate extension programmes. Initiate activities for developing Strategic Research and Extension Plans in all the districts to be followed up with Block level plans.
- Improve the capabilities of extension managers-Extension managers need skills to operate effectively in the pluralistic extension environment. The main challenge for an extension manager would be managing the process of change in extension. A series of training programmes have to be initiated to provide these skills.

Future

The role of agricultural extension in the next decade should be quite different from what it was 10 years ago or even now. Its role as a facilitator of agricultural knowledge system would only increase as more participants from private sector would get involved in extension. It is likely that input related extension (seed, fertiliser, machines and chemicals) would move to private sector in the future.

Escalation of Agricultural Extension System in India

Agricultural extension in general terms means educating farmers and ensuring their participation in application of scientific research and research based knowledge to field situations for wider acceptance and subsequent dissemination of superior farming technologies. The field of 'extension' encompasses a wider range of communication and learning activities organized for agri-preneurs including farmers, farm women and rural youth. The process of agricultural extension although started way back in 1800 B.C., it was not organized and structured worldwide until the late twentieth century. In India, Agricultural extension started with individual efforts of some philanthropists during the 1860s and moulded through situational needs, urgencies socio-politico decisions and research based remodelling in few cases.

Post Independence Reforms in Agricultural Extension in India

Gradual Evolution of Public Extension Strategies

The extension strategies followed by the public sector extension agencies to extend their outreach to larger number of agri-preneurs evolved over time. Among the important public extension strategies witnessed since the last few decades in India, following are some notable ones:

- Multi-agency involvement and pluralistic mode of extension delivery, through KVKs, ATMA, Central/ State government schemes, specialized programmes etc.
- Farming system approach and Strategic Research and Extension Plan (SREP) as implemented through the KVKs and ATMA.
- Farmer-centric extension services.

- Research–extension interface and convergence.
- Gender budgeting and gender mainstreaming.
- Mass media and HRD support.

Given the diversity of the country, nature of the economy and its limitations, there is a need for better co-ordination and complementary role playing among public and private agencies as well as NGOs to promote pluralism and partnership. Although in a country like India, it is hardly possible to completely bring into effect fee-based extension services, there is a need for the district level public extension organizations to raise funds through practically implacable means.

Agricultural Extension Approaches in India

For a long time since independence, extension services were being provided mainly by the public sector. At present public sector is major extension service providers through a two-tier system. At the central level, Indian Council of Agriculture Research (ICAR) is the nodal institute for agriculture research and extension; while at the state level, the State Agricultural Universities (SAU) via the Krishi Vigyan Kendra (KVKs) and Agriculture Technology Management Agency (ATMA) at the district level facilitate agriculture extension. However, the public extension is highly skewed towards crop husbandry ignoring allied sectors. Besides the existing public extension service system, there are several private players, civil-society organizations including farmer-based organizations and NGOs that play a major role in providing extension services.

Agriculture Extension System by Non-public Players

Input Dealers

At present, there are 2.82 lakhs input dealers in the country who are marketing seeds, pesticides, fertilizers and farm machinery. A study estimates that 13% of farmers get their information from input dealers in the country. However, a major complaint against input dealers is that they involve in product promotion instead of technical advice to the farmers. To address this issue, the government of India recognizing the importance of this category of extension support to farmers has started offering a course by MANAGE and other recognized institutes on input use efficiently this will help Input Dealers to brush up on the latest technical knowledge in various sub-sectors of agriculture and to certify them as qualified Input Dealers.

Agribusiness Companies

Private sector agribusiness and input manufacturing companies (ITC, Nuziveedu Seeds, IFFCO, KRIBCO, Pepsico, BAYER, Pioneer, Monsanto etc.) also undertake direct extension activities, these extension activities are in support of their product brand and seek to help the farmers to realize higher production (and thus returns) through necessary pre-sowing preparation, optimum seed rate, correct agronomic practices, application of nutrients, harvesting techniques and plant protection measures.

NGOs

In India, there are 3 million registered NGOs of which many are actively involved in the development of rural areas and naturally oriented themselves towards land-based livelihoods. Hence, they accommodate an essential component of extension in their intervention. PRADAN, BAIF, Syngenta Foundation, India etc. are some of the leading NGOs working in agriculture and allied sector.

Organized User Groups

Farmer Producer Groups, Commodity Interest Groups, Farmers Clubs, Farmer's Cooperatives etc. roles in extension in niche regions and areas. Some of the most successful farmer organization in Maharashtra is MAHAGRAPE, and United Planters Association of Southern India (UPASI) which is the apex body of tea, coffee, rubber and cardamom growers is involved in leadership, research and extension services for the plantation industry. Generally, farmers groups play an important role in reducing the cost of cultivation, increased profitability, information access and reduce the transaction cost there by facilitating farmers to achieve higher price realization. The consumers also get benefited as they get good quality products at lesser cost. Therefore, many states and government of India promoting Farmers Producers' Organizations (FPOs) as they also play important role in strengthening extension activities. For example farmer's cooperatives are playing an important role in the application and dissemination of new technologies. Cooperatives collaborate with research institutes and extension organizations, and provide services to their members. At present more than 95% of coóperatives provide effective technology services and help to solve the problems. Majority of the farmers' cooperative organizations provides its members with seeds, pesticides, fertilizers and farm machineries through group buying and direct purchase and this will reduce the price by 10% and increases the profit by 25-30 per cent.

Extension through Media and ICT

ICT is an imperative tool to link value chains in agriculture. Due to the increasing number of ICTs tools such as mobile phones, internet, TV etc. has shown tremendous potential to disseminate information to farmers. Kisan Call centres have proved to be important and reliable arm disseminating information and addressing farmer's queries. Every week more than 1.9 crore crop advisories reach out to farmers. Mobile Apps, TV channels and periodic journals have benefitted e-friendly farmers in a massive way.

Present Status of Agriculture Extension in India

Agricultural extension in India is skewed towards crop production neglecting allied sectors, though the allied sectors such as dairy and fishery significantly contribute towards farmer's household income. However, extension support is weak in the case of animal husbandry and fisheries in India. The public sector extension is currently offering a one-size-fits-all to all category farmers and all kind of products. Considering heterogeneity among farmers, crops they produce, agro-climatic zones, soil types, farm resources endowments, enabling policy environment, etc., there is a need for continues capacity building of extension professionals as they lack recent advancements in agriculture technologies and development. In the case of NGO led extension models are relatively efficient and exhibit sensitivity to local priorities, however, they lack capacity and scale to make to create impact at large. It is evident that non-public extension players offering extension services in a localized region without any coordination among the players and even with the public extension. As a result, there is a restriction of good practices developed among individual players and an opportunity to scale applications at large. To mention, the majority of the players operate without any quality standards, as a result players are unaccountable leaving farmers alone towards legal issues under the circumstances of harmful advice (it happen mainly with seed business). The right type of information makes a huge difference to farmers, inappropriate and untimely information has serious consequences on his farm decisions and profitability. The negative impact of imperfect knowledge has proven to be a major problem especially in the private sector. Spurious markets for pesticides and fertilizers exist in different parts of the country.

■■■■

Chapter - 2

New Trends in Agriculture Extension

1. CYBER EXTENSION

Cyber extension is an agricultural information exchange mechanism over cyber space, the imaginary space behind the interconnected computer networks through telecommunication means. It utilizes the power of networks, computer communications and interactive multimedia to facilitate information sharing mechanism.

Cyber extension is an effective and viable tool of dissemination of information not only for farmers but also for students, extension personnel, scientists, policy makers, planners and other stake holders. It is an integration of various technologies using network and the process of communicating the desired and the needy information to the target clients for making them more participative and independent in their decision making. Cyber extension is cheap, often updated and can reach maximum number of targets at a minimum time and more importantly it is an interactive media. Cyber extension will bring new information services to the rural areas which farmers, as users will have much greater control over current information channels. Several approaches like 'Warana Wired Village Project (Maharashtra)', 'Cyber Grameen (Swarna Bharat Trust)', E-Chaupal (ITC), I-Kisan (Nagarjuna Gorup of Companies) etc have been tested and doing well.

The cyber extension tools and techniques affect the whole extension system in a big way as it reduces the dependency for personal advice to a great extent. In most case cyber extension will not even replace the traditional extension. Rather it will add more interactivity and adds to two way of communication. Cyber is the most flexible medium of communication currently available. Kisan Call Centres: Toll Free Telephone Number 1800-180-1551 considering the impressive telecom network in the country, Kisan Call Centre scheme started by the Government of India, Ministry of Agriculture, which is based on extension service delivering knowledge and information exactly as per the requirements of the farming community by using telecom network. The scheme has been functioning from 2nd January 2004. The call centres can be accessed by farmers all over the country on common Toll Free Telephone Number 1551 which can be dialled from anywhere in the country, the location is immaterial as the call can originate from any village to land at a specific call centres.

Definitions of cyber extension

Cyber extension is defined as the 'extension over cyber space'. But, in applied context of agriculture, cyber extension means, "using the power of online networks, computer communications and digital interactive multi-media to facilitate dissemination of agricultural Technology". It includes effective use of ICT, national and international information networks, internet, expert systems, multi-media learning systems and computer based training systems to improve information access to the farmers, extension workers, research scientists and extension managers.

Cyber Extension can be defined as the information is an important resources in modern agriculture. The development of computers and improvements in telecommunication offers farmers and extension workers, many now opportunities to obtain technical and economic information quickly and use it effectively for their decision making.

Thus cyber extension is "Extension over virtual space". Cyber extension includes effective use of information and communication technologies, national and international information, Networks, Internet, Expert systems, Multimedia learning systems and computer based training systems to improve information access to the farmers, extension workers, research scientists and extension managers.

But in the applied context of agriculture, cyber extension means "using the power of online net works, computer communication and digital interactive multimedia to facilitate dissemination of agricultural technologies".

Objectives of cyber extension

There are following objectives of cyber extension:

1. To create awareness among extension functionaries.
2. To orient the extension managers and researchers on the need and advantages of using information and communication technology.
3. To find out effectiveness of use of cyber extension as ICT media.
4. To discuss and appreciate the continued importance of e-mail, Telnet, FTP, Gopher, www for wider reach of agricultural information.

Concept of cyber extension

- Cyber space the imaginary or virtual space of computers connected with each other on Networks, across the world.
- These computers can access information of Text, Graphics, Audio, Video and Animation files. Software tools on networks provide facilitates to interactively access the information from connected servers.

Cyber extension in agriculture and rural development

The utilization of agricultural and rural cyber extension is very useful for rural farmers to increase crop yields, as well as to consult researchers about agricultural issues. In some countries, it serves to escalate society's learning. By utilizing online networking, computer and digital interactive multimedia, not only researchers and farmers, but also by the whole society can benefit from it.

Cyber extension is regarded as the use of ICTs for agricultural and rural extension. There are three systems developed of cyber extension which are:

i. Online consultation system -web consultation.

ii. Hand phone -based internet system.

iii. Android technology -based smart telephone.

The utilization of agricultural and rural cyber extension is very useful for rural farmers to increase crop yields, as well as to consult researchers about agricultural issues. In some countries, it serves to escalate society's learning. By utilizing online networking, computer and digital interactive multimedia, not only researchers and farmers, but also by the whole society can benefit from it. Cyber extension is regarded as a strategic model since the model could improve information accessibility for farmers, field extension officers, extension managers, researchers, input agents and other related fields on extension practices. The utilization of cyber agricultural and rural extension is not only performed by developed countries but also by developing ones.

Role of Cyber Extension in agricultural development

Agricultural extension played a very important role in achieving the targets of green revolution. The post green revolution scenario is witnessing emergence of new paradigms like sustainable agriculture, agri-business and contract farming. The advances in information communication technology (ICT) are poised to compliment extension efforts for transfer of technology.

Milestones in Cyber Extension

During 1995, ARIS (Agricultural Research Information System) was established to help students and farmers to get immediate information regarding agriculture. During 1997, agrisnet was started by GOI. Then, during 1999, ATIC was started under NATP project. Pilot projects to connect rural community to the cyber space are underway at various locations. MANAGE has established internet connectivity in 28 districts of 7 states *viz.*, AP, Bihar, HP, Jharkhand, Maharashtra, Orissa and Punjab under NATP. MANAGE has taken the initiative to provide linkages to the technical and other farmer friendly information through its web-site. It is also supporting a number of AU's and other research and training organizations, both in public, voluntary sector in building their capacity to digitize the agriculture. Information and host the same on the web. Then many attempts were made in the field of cyber extension. Some are:

1. Information village project.
2. Warana wired village project.
3. E-choupal.
4. Kisan call centre.
5. Recently, Mission-2007 was attempted which is also called as 'Knowledge village centre'.

Information Village Project

This project was launched during 30 Dec, 1999 by the M S Swaminathan Research Foundation. This project linked ten villages near Pondicherry with computers. These new technology tools are empowering everyone with knowledge and opportunity by an inclusive use of local languages and a multimedia format that allows all to participate. Information provided in the VKC is local specific and relates to prices of agricultural inputs (such as seeds, fertilizers, pesticides) and outputs (rice, vegetables), market (potential for export), entitlement (the multitude of schemes of the central and state governments, banks), health care (availability of doctors and paramedicals in nearby hospitals, women's diseases), cattle diseases, transport (road conditions, cancellation of bus trips), weather (appropriate time for sowing, areas of abundant fish catch, wave heights in the sea), etc.

Warna Wired Village Project

Actual implementation is done in April 1998 by GOI through National Informatics Centre (NIC), Government of Maharashtra and Warna Co-operative Society with the share of financial support being in the ratio of 50:40:10. This project provides information on different crop cultivation practices of major crops, sugarcane cultivation practices, pest and disease control, marketing information, dairy and sugarcane processing information etc. to the farmers, right up to the village level.

E-Choupal

This was started during 2004 by ITC, Madhya Pradesh. 'e-Choupal' unshackles the potential of Indian farmer who has been trapped in a vicious cycle of low risk taking ability - low investment - low productivity - weak market orientation - low value addition - low margin - low risk taking ability. Such a market-led business model can enhance the competitiveness of Indian agriculture and trigger a virtuous cycle of higher productivity, higher incomes, enlarged capacity for farmer risk management, larger investments and higher quality and productivity.

Kisan Call Centres

The Department of Agriculture and Cooperation, Ministry of Agriculture, Govt. of India has launched Kisan Call Centres during 2004 with a view to leverage the extensive telecom infrastructure in the country to deliver extension services to the farming community. The aim of Kisan Call Centres to respond to issues raised by farmers instantly in the local language, on continuous basis. A Kisan Call Centre consists of a complex of telecommunication infrastructure, computer support and human resources organized to manage effectively and efficiently the queries raised by farmers instantly in the local language.

Kisan call centres operating in India

Location /States covered by Kisan Call Centre

1. Mumbai (Maharashtra) and Goa, Daman, Diu.
2. Kanpur (Uttar Pradesh) and Uttaranchal.

3. Kochi (Kerala) and Lakshadweep.
4. Bangalore (Karnataka).
5. Chennai (T.N.) and Andaman/ Nicobar.
6. Hyderabad (Andhra Pradesh).
7. Chandigarh, J&K, Himachal Pradesh and Punjab.
8. Jaipur (Rajasthan).
9. Indore (Madhya Pradesh) and Chhattisgarh.
10. West Bengal, Bihar, Orissa and Jharkhand.
11. Kolkata and N.E. States.
12. Delhi and Haryana.
13. Ahmadabad, Gujarat and Dadra & Nagar Haveli.

Apart from the above mentioned lists some other attempts were made in the field of cyber extension. During 2007, M S Swaminathan introduced MISSION-2007 (Knowledge Village Centre). PURA (Provision of Urban Amenities in Rural Area) was also implemented by GOI. To make all the above attempts successful, Government of India also started Computer Literacy Mission.

Future scope of cyber extension

Cyber extension is gaining importance in India. There are as:

- Informations given are not in the local language.
- Some informations given may not hold good for local conditions.
- Illiterate farmers cannot get the benefit of cyber extension.
- Dissemination of technology through Cyber Extension in highly remote areas is a difficult task.

Advantages of cyber extension

Cyber extension methods are way ahead in efficiency than other ordinary extension methods. But there are advantages and drawbacks of cyber extension systems. Following are the advantages of cyber extension:

i. It is the most revolutionary method of agricultural extension teaching methods.
ii. Due to its convenience, the latest technologies invented in the other end of the world will reach in a minute to the local farmers.
iii. Rural communities do not feel "we are isolated", instead of that they feel "we are connected to the world".
iv. Farmers become computer-literate.

v. The Internet will become a rich source of education not only for the farmer, also for his/her family.

vi. Farmers will feel good about their personality when using cyber extension; they feel honoured.

vii. Free access to the websites will increase the farmer participation in extension programs.

viii. Farmers will be well educated and aware of agricultural technologies by seeing, hearing and watching the demonstrations.

ix. Farmers will be well informed of latest issues related to agriculture around the world.

x. Due to cyber extension systems up to date data, farmers are able to forecast the weather, their harvest and even market trends.

xi. Reading newsletters on the websites, farmers do not face the drastic situations such as price fix, price fluctuations, intermediaries smuggling, black market prices etc.

xii. Unlike traditional farming, farmers can seek new ventures for their excess production (starting jam, cordial production, investing in mushroom cultivation, compost production, milk and meat processing etc.).

xiii. Due to technological advancements, the young generation tend to involve in agricultural activities.

xiv. Extension officers can deal with extension programs without worrying about the other office work, time and transportation costs etc.

xv. Extension officers can pre-record their demonstrations and deliver it later to the farmer community.

xvi. Unlike the traditional lazy extension officials, use of cyber extension will make them actively participated in the extension programs.

xvii. The dynamic nature of extension programs will be increased.

xviii. The extension officers will have to work less effortlessly (as farmers know new methods due to cyber extension) due to pre-awareness of the farmers.

xix. Extension officers can follow up the work; create progress reports done by farmers online without reaching them.

xx. Extension officers can change extension programmes according to its compatibility (situational –new varieties of paddy to be cultivated in a different agro-ecological zone, cultural - website of Agricultural Department of Sri Lanka is trilingual).

xxi. Problems and risks related to trialabilty is reduced to a major extend.

xxii. Observability of agricultural aspects not only in local, also in international level is open to general public.

xxiii. Extension programs will be no longer extensive; it will be close and easy as just as a click of a mouse.

Drawback of cyber extension

Following are the drawbacks of cyber extension

i. As cyber extension facilities are not available in every village, farmers may have to travel far for it.

ii. Due to rapid changes occur in technology, farmers will be left with a doubt and they will find it difficult overcome the situation of "information overload".

iii. Most of the rural farmers are still illiterate. For such people it will be harder to become computer literate.

iv. As still a developing country, seeing the latest agricultural advances in developed countries (e.g., Canada) will leave the farmers and the general public a feeling of shame (like they have it all, we have nothing), a doubt within themselves.

v. The cyber extension will loosen the bonds of a family (the children will consume their time logged in facebook, etc. rather than helping the farming).

vi. Free access to the internet will degrade the quality of the cyber extension.

vii. False information, fake facts will be a great threat not only to the agricultural activities but for the general awareness also (e.g., one website may tell that using heavy metals in minimal amounts in the fertilizers may not be a risk for your health, and in another website you may see the traces of heavy metals is lethal).

viii. As the cyber extension is available in Agrarian Service Centres during the office hours, the farmer will have to give up farming operations during the daytime.

ix. New ventures illustrated in cyber space are not modified into locality. Adaptation of such technology will risk the harvest. Some of the materials used in foreign cyber extension networks, are not available in our country.

x. If there is a malfunction in the cyber extension system, there is a risk of losing all the data kept for years. A back-up plan should always be available.

xi. As there are thousands of information available in cyber extension, extension officers may include false data in their reports (Like copying and pasting from another website, unaware of what to bed one).

xii. Most of the cyber extension systems do not issue their data free. They require endless registrations, sign-ups, passwords, private details etc.

xiii. Most of the cyber extension systems are not situational and cultural compatible.

xiv. Information provided is not in local languages.

xv. Dissemination of technology through cyber extension in highly remote areas, extreme corners of the country is a difficult, unattainable task.

xvi. Observability of data is banned in certain regions of the world.

2. MARKET - LED EXTENSION

- Market – led extension as a market ward orientation of agriculture through extension, include agriculture and economics, an is the perfect blend for reaching at the door steps of farming community with the help of appropriate technology – Khaleel, 2007.
- Market – led is the identification of customer needs and wants before offering a service.
- With globalization of the market, farmers need to transform themselves from more producers – sellers in the domestic market to producers cum sellers in a wider market.

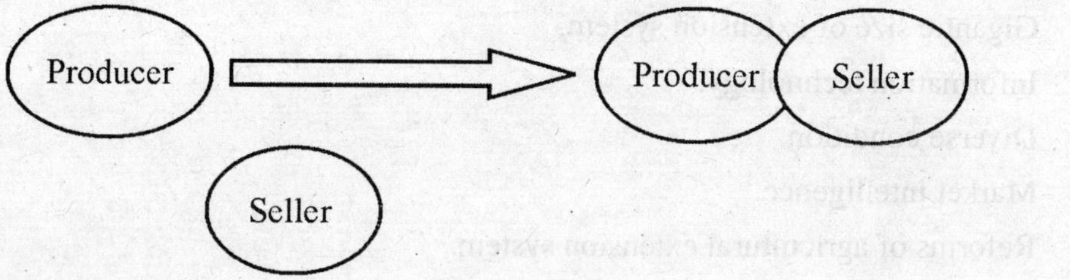

Market – led extension is

- market ward orientation of agriculture through extension.
- include agriculture and economics.
- The perfect blend for reaching at the door steps of farming community with the help of appropriate technology.

Element of market – led extension

- Need based.
- Demand driven.
- Emphasis on efficiency.
- Client orientation.
- Aims at high return to farmers.

Roles of agricultural extension personnel in Market – led extension

- SWOT analysis of the market.
- Organization of Farmers Interest Groups (FIGs).
- Enhancing the interactive and communication skills of he farmers.
- Establishing marketing and ago-processing linkage.
- Advice on product planning.
- Educating the farming community.
- Regular usage of internet facility and production of video films of success stories of commodity specific farmers.

- Direct marketing: farmers need to be informed about the benefits of direct marketing.
- Acquiring complete market intelligence.
- Publication of agricultural market information.
- Capacity building of FIG in terms of improved production, post harvest operations, storage, transport and marketing.

Challenges to market – led extension

- Gigantic size of extension system.
- Information technology.
- Diverse condition.
- Market intelligence.
- Reforms of agricultural extension system.
- Market size is large and continuously expanding, but marketing system has not kept pace.
- Direct marketing "farmer – consumer" is negligible.
- 85% of the 27294 rural periodic markets, facilities for efficient trade is still almost absent.
- 7200 market yards/subyards are inadequate, ill equipped and mismanaged.
- Due to lack of proper handling at farm gate lead to 30% fruits and vegetables, 7% grains, 10% spices loss before reaching market.
- Rs 5000 crores/year lost due to poor marketing chain.
- Risk bearing.
- Storage of farm produce and grading.

Different models of MLE in India

Rythu bazaar in Andhra Pradesh	Apni Mandi in Punjab	Uzhavar Santhi in Tamilnadu	Shelkari Bazar in Maharatra	Raithara anthe in Karnataka	Krushuk Bazar in Orissa

Philosophy and principles of MLF

- Farmers and local institutions (e.g. producer organizations or village leaders) should play a key role in selecting farmer –trainers and monitoring and evaluating them. This helps make the programmes more accountable to the community or groups that they serve.

- Farmer-trainers are "of the community": they communicate in local languages and more sensitive to local cultures, mannerisms, farming practices and farmers' needs.
- Farmer-trainers should be selected on the basis of their skills and interest in sharing information, not just on their farming expertise.
- Farmers-trainers need strong linkage with and support from development agents (whether government, non-government organization (NGO or private), the people who train and backstop them. Farmer-trainers generally serve as a complement to exiting extension systems, rather than being a substitute or them.
- Facilitating organizations and local institutions need to be proactive in ensuring that women as well as men become farmer-trainers.
- Simple and appropriate reference materials should be made available to the farmer-trainer.

Problems in MLE

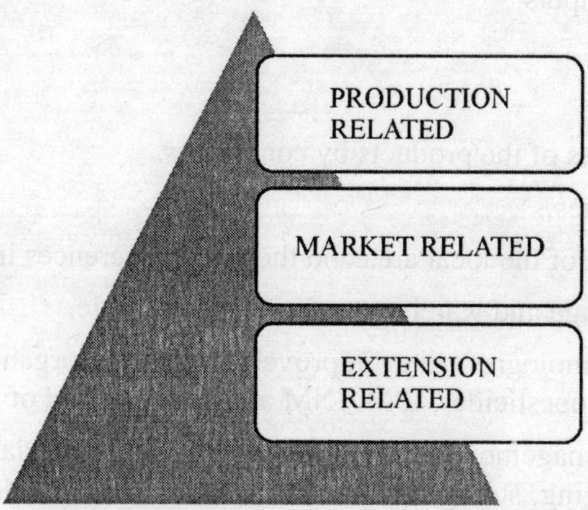

PRODUCTION RELATED	MARKET RELATED	EXTENSION RELATED
• Seasonality of production: supply not uniform throughout the year • Parishability of produce: Problems of storage • Bulkiness of production: Transportation problem	• Non -availability of market information • Existence of many middlemen • Inferior quality of produce • Export promotion • Supply chain management • Retail market	• Lack of communication skills • Lack of credibility • Insufficient information related with market and many more

Objectives of MLF

- Conversion of agriculture sector in to profit oriented business.
- Strengthening R-E-F linkage in between various department at various levels.
- Strengthening market linkages to farmers – IT application in agricultural marketing.
- Wider use of electronic mass media for agricultural extension.

Required information to extension system and farmers for MLE

- Present agricultural scenario and land use pattern.
- Suitability of land holding to various crops and enterprises.
- Crops in demand in near future.
- Market prices of crops.
- Availability of inputs.
- Usage of inputs.
- Credit facilities.
- Desired quantities of the products by consumers.
- Transport facilities.
- Market net work of the local area and the price differences in various markets.
- Network of storage and warehouses facilities available.
- Production technologies like improved varieties, organic farming, usage of bio fertilizers and biopesticides, IPM, INM and right method of harvesting etc.
- Post harvest management like processing, grading, standardization of produce, value addition, packaging, storage, certification, etc. with reference to food grains, fruits and vegetables, eggs, poultry, fish etc.
- Contact farming.
- Private modern terminal markets.
- Food retail chains.
- Food safety and quality standards.
- Certification.
- WTO regulations.

3. FARMER-LED RESEARCH AND EXTENSION

- Farmer-led research and extension (also called "participatory technology development, or PTD) combines the knowledge and research capacities of local communities and research and development organizations in an interactive learning process. It involves identifying, generating, testing and adapting new techniques and practices to help solve local problems.

- The ultimate aim is to strengthen the experimental and technology management capacities of local people and communities, thus farmers play a key role in the entire process. The "P.' in PTD can also refer to "people-centred" strategies and processes.
- Farmer – led extension/farmer to farmer extension (F2F) is defined as " the provision of training to farmers to farmers, often through the creation of a structure of farmer promoters and farmer trainers" (Scarborough et al., 1997).
- We use 'farmer trainer' as a generic term, even though we recognize that different names (e.g. lead farmer, farmer promoter, Community knowledge worker) may imply different roles.
- F2F extension programs date back atleast to the 1950, currently such programs are widespread.

Principles of farmers – led Extension

- Consider farmers' needs, indigenous knowledge, existing resources and local networks. Foster the development of these resources.
- Gain joint understanding of the main characteristics of and changes in the agroecological system.
- Support farmers and their organizations to increase their awareness, self-respect, self-confidence, knowledge and skills. This also encourages local people to continue the research and extension process after outside facilitation ceases.
- Ensure that farmers and outside facilitators jointly define priority problems.
- Provide practical experience to selected farmers so they can develop new technological options and transfer their knowledge to others.
- Use low-cost, locally produced research and extension inputs and materials. Let farmers and their organizations circulate them. This ensures that farmers are self-reliant and inputs are appropriate.
- Encourage demonstrations by a farmer or farmer group on their farm(s). The demonstrations can be replicated by other farmers.
- Promote research and extension roles for farmers themselves. Farmers traditionally fill these functions and they should not be taken up by outsiders who often know less about local physical and sociocultural conditions.
- Provide information on the changing situation to create awareness.
- Experiment locally with a variety of options derived both from farmers (indigenous knowledge or other farmer experiences) and from formal science. Suggest optional practices for farmers to decide on and test in their own fields, and encourage farmers also to suggest practices to test.
- Hold fore for farmers to evaluate and extend the research results to others.

Farmer-led vs. conventional research and extension

- Farmer-led approaches are complementary to (not a substitute for) station-based research, scientist-managed on-farm trials and conventional extension approaches.
- Farmer-led approaches link the power and capacities of agricultural science to the priorities and capabilities of farming communities.
- Farmer-led and conventional research and extension activities differ in three essential ways.
- Farmer-led approaches do not attempt to generate results that can (or should) be generalized across wide areas (although that may happen).
- Farmer-led methods follow a different approach to collecting, codifying, interpreting and utilizing information.
- Practitioners of farmer-led approaches can be based in any rural development service or project and even among a local community.

Limitations of farmer - led research and extension

- Farmer experiments are often undirected and unfocused. This is a result of (or leads to) poor experimental design.
- The relevance and application of research can be limited.
- Political friction, language problems, remoteness and inaccessibility can limit the ability of researchers and extension workers to follow up experiments and of local farmers or communities to share their experiences.

Lead farmers

An effective lead farmer is the one who always produces the best crop in his/her field, takes up new innovations as quick as possible and is willing to train other farmers. The person should not hold another leadership role in the community because work will be sacrificed for other responsibilities. The field of the Lead farmer becomes a learning centre, where other farmers can come and learn. The lead farmer should also be accepted by the farmers he/she mentors to be effective.

Criteria for selection of lead farmers

- Farming expertise/hand working can be role model.
- Literacy: able to read and write.
- Residence in the community.
- Able to communicate the things.
- Good behaviour, trustworthy, acceptable to community
- Easy reachable/available.

- Good track record (not defaulted on loan in past, no criminal record).
- Innovative.

Farmer Field School

- Farmer Field School (FFS) consist of groups of farmers who get together to study a particular topic.
- The topic covered vary from conservation agriculture, organic agriculture, animal husbandry and soil boundary, to income generating activities such as handicrafts.
- FFS provide opportunities for learning by doing. It teaches basic agricultural and management skills that made farmers experts in their own farms.
- FFS is a forum where farmers and trainers debate observation, experiences and present new information from outside the community.

Thus,

FFS creates awareness, supports learning, transfer technology and assist communities in problem solving.

Basic concepts of FFS

If I listen, I may forget;

If I see, I may remember;

If I do, I may remember;

If I discover, I may own it;

If I practice, I may perfect;

In FFS, one can see, do, discover and practice.

Essential elements of FFS

- The group
- The field
- The facilitator
- The curriculum
- Programme leader
- Financing

Goal/objectives of FFS

- The goal of FFS is mobilization, sensitization and empowerment of rural farming communities.

- In small group (20-30 people) they identify their production gaps, test for themselves best on-farm options and practices which are adaptable and sustainable to their farming situation.

Special features of FFS

- All learning is field based and it is primary venue for learning.
- FFS is group learning constantly over the experimentation period.
- FFS learning promotes healthy decisions and quality decisions.
- Farmers conduct their own field studies with comparison or treatments.
- Learning continue until a full crop season.
- Facilitates farmer to farmer communication.
- Field staff serve as facilitators.
- Farmer field schools is a unique way to educate farmers.
- It is an effective platform for sharing of experiences and collectively solving agricultural related problems.
- Since 1989 more than two million farmers across Asia have participated in this type of learning.

Benefits of Farmers Field School

i. The FFS can facilitate technology transfer among farmers.

ii. The FFS is stills a new confidence among participants to understand and sustain their agro-ecological environments.

iii. The FFS brings farming communities together in a unique and productive manner. It improves relationships between farmers and researchers.

iv. The FFS is an advance on the previous doctrine of the Farming Systems Research and Extension Methodology.

Farm School

- "Farm school are established in the field of such innovative farmers or farmer achieves who are actually enhancing productivity and profitability in their farms through scientific and sustainable agriculture."
- Farm school is powerful instrument for participatory research and knowledge management.
- It is an alternative extension tool and facilitate farmer –to – farmer learning.
- It also reduces the widening gap between scientific know-how and farmers practices.
- This would facilitate development of fa4rrming situation/system specific packages.

Concept of farm school

i. Farm Schools would be operationalized at Block/Gram Panchayat level.

ii. These would be set up in the field of outstanding or achiever farmers. The list of such farmers having potential for organising farm schools in the block for different sectors will be identified jointly by FAC and BTT and sent to ATMA GB through the Panchayat Samiti (block level elected body of Panchayati Raj Institutions). "Farm Schools" & "Farm School Trainers" for each season will finally be approved by the ATMA GB.

iii. "Teachers" in the Farm Schools could be progressive farmers, extension functionaries or experts belonging to Government or Non-Government Sector.

iv. One of the main activities of Farm Schools would be to operationalize Front Line Demonstrations in one or more crops and/or allied sector activities. These demonstrations would focus on Integrated Crop Management including field preparation, seed treatment, IPM, INM, etc.

v. Farm Schools would provide season long technical backstopping/ training to target farmers by having an interactive session once at least during each of the 6 critical stages in a cropping season.

vi. While selecting the trainee farmers, about 50% representation should be given to small and marginal farmers. Preference may be given to members of CIGs / FIGs.

vii. "Students" will visit Farm Schools as per specified schedule or as may be necessary. "Teachers" may also visit students as may be necessary.

viii. Knowledge and skills of "teachers" would be upgraded on a continuous basis through training at district/state/national level institutions and/or exposure visits, etc.

ix. In addition to technical support through Farm Schools, knowledge and skill of "students" may also be upgraded through training at district/state level and exposure visits, etc.

"Students" would have the responsibility of providing extension support to other farmers in the respective village or neighbouring villages.

4. EXPERT SYSTEM FOR AGRICULTURE EXTENSION

Expert system can be defined as a tool for information generation from knowledge. Information is either found in various forms or generated from data and/or knowledge. Text, images, video, audio are forms of media on which information can be found, and the role of information technology is to invent, and devise tools to store and retrieve this information. Statistical information is a good example of information generated from data while advises generated by an expert system is a good example of information generated from knowledge. At the beginning, the concentration was mainly on textual information. Information technology used for textual information ranged from paper archiving tool to the sophisticated electronic computer software tools such as data base management systems, and hypertext tools. Images were very difficult to be included in an information system just few years ago till the scanners technology has been advanced together with the invention of the optical storage devises which have the capacity to store hundreds of mega bytes. This technology has enabled the developers to include images in their systems. The need of expert systems for technical information transfer in agriculture can be identified by recognizing the problems in using the traditional system for technical information transfer, and by proving that expert systems can help to overcome the problems addressed, and are feasible to be developed.

Information Transfer Problems

- **Static Information:** Examining the information stored and available in the agriculture domain revealed that this information is static and may not respond to the growers need. All extension documentations give general recommendations because there are many factors if taken into consideration; so many different recommendations should be included in the document.
- **Specialties Integration:** Most of the extension documents handle problems related to certain specialty like plant pathology, entomology, nutrition, or any other specialty. In real situations, the problem may be due to more than one cause, and may need the integration of the knowledge behind the information included in the different extension documents and books.
- **Combination of more than one information source:** Images may need sometimes an expert to combine other factors to reach an accurate diagnosis, and even if a diagnosis is reached, the treatment of the diagnosed disorder should be provided through extension document.
- **Updating:** Changes in chemicals, their doses, and their effect on the environment should be considered. Updating this information in documents and distribute them takes long time. The same arguments can be made for audio tapes that are another form of extension documents but in voice instead of written words. Video tapes are more stable than other media as the information provided through the tape describes usually well established agricultural operations. However, if the tape includes information as what is commonly included in documents and audio tapes, this information should be updated.
- **Information unavailability:** These are cases, where information may not be available in any form of media. It can be obtained only from human experts, extensionists, and/or experienced growers. In addition, the information transfer from specialists and scientists to extensionist and farmers, represents a bottle neck for the development of agriculture on the national level. The current era is witnessing a vast development in all fields of agriculture. Therefore, there is a need to transfer the information of experts in certain expert system for agriculture extension

Domain Application Aspects: The agriculture domain can be classified into subdomains namely: plant production, animal production and management of natural resources related to the agricultural operations such as soil and water. Expert system has been applied in the three subdomains. Expert systems for field crops are implemented for : diagnosis of soybean diseases, crop management for cotton and weed identification for wheat. Expert systems were also implemented for horticulture crops: apple orchid management, cucumber production management. Agroforestry is another area in plant production where expert systems have been developed. Some other applications cannot be categorized commodity wise such as the expert system developed for selecting evapotranspiration estimation methods. Another way for classifying agricultural expert systems is the domain specific task that this system performs such as: irrigation, fertilization, pest management, diagnosis of plant diseases, and others. There are mapping between the domain specific tasks and the generic tasks. For example irrigation application is mainly a special case of scheduling whereas diagnosis is a special case of classification.

Ontology Based Expert System: In earlier systems, a vast amount of knowledge was stored in the knowledgebase as rules. Rules were extracted manually and stored in the expert system. In our OBES, domain knowledge is stored in ontology, which is easy to create using available GUI editors. OBES have the advantage of distributive development and maintaining of knowledgebase over traditional expert systems. The system allows users to classify the crop ontology and checking its consistency.

Knowledge sharing and reuse: Knowledge sharing and reuse is one of the topics that have attracted the attention of the AI researchers in the last few years. The research in knowledge sharing in agriculture can be directed toward identifying common knowledge that can be shared among different expert systems such as identification of agriculture ontology, knowledge related to common resources namely: soil, water and climate, knowledge related to the same taxonomic category of a set of crops, etc. The research in knowledge reuse can be directed to building a library of domain specific tasks in agriculture such as: irrigation, fertilization, integrated pest management, etc.

Intelligent retrieval of agricultural data: Meteorological data are very important for agronomists as forecasting weather data helps in giving recommendations to growers. A system that serves as an intelligent assistant for meteorologists to locate and analyze historical situations of interest, has been developed using case based reasoning (Jones and Roydhouse, 1995). This work has been oriented to be used by meteorologists but not agronomists. The basic idea could be investigated to retrieve historical situations that are important for managing different crops.

Multimedia and expert systems: Integrating multimedia with expert systems is a hot topic that is booming now-a-days. The integration with images was frequently done to more efficiently acquire the user inputs, whereas other types of media such as sound, and video are also addressed. An example of integrating multimedia with expert systems. It was found that describing symptoms in words is very difficult and sometimes is very confusing. Therefore, images are identified to be used for two main purposes: describing a disorder symptom, and confirming the diagnosis of the cause of a certain disorder. Detailed images for all symptoms, and unique images that confirm the occurrence of disorders at different stages should be collected. Although images are very useful in acquiring the user inputs, the uncertainty problem is still there. Therefore, giving the user the option to select an image with a degree of certainty should be provided. Providing more than one picture for the same symptom can reduce the user uncertainty, but this will lead to exerting more efforts in collecting and classifying the images. As the output of an expert system for crop production management, is a set of agricultural operations, describing how to perform an agricultural operation in words, is very hard and one can never guarantee that the user can understand what has been written. Displaying a video for a professional doing the recommended operation would be very educational. The sound is essential because sometimes, it is not easy to write terminology used by growers in daily life. In addition, combining the video with sound is also recommended to comment on how the operation is done. Although the given example proved the possibility of integrating multimedia with expert systems, there are still some problems which needs further research such as the intelligent selection of the appropriate media for presentation taking into consideration the user level, getting input data from images, for example providing the expert system with an infected leaf of a plant for diagnosis, and/or enhancements of all input devices interfaces, which is a general research issue, in order to provide expert system with data in different forms.

5. PRIVATIZATION IN EXTENSION

- Savas (1987) defined privatization as: An act of reducing the role of government or increasing the role of private sector in an activity or in the ownership of assets.
- Process of funding and delivering the extension services by private individual or organization is called private extension.
- Privatization is the act of reducing the role of government or increasing the role of private sector in an activity or in the ownership of assets.
- Privatization in the incidence or process of transferring ownership of business from the public sector (government) to the private sector (business).
- Private extension is solely the act of private individuals or organizations where the decision of privatization solely rests with government implemented in liaison with Private Extension Service Provider services rendered in the area of agriculture˙ and allied aspects by extension personnel working in private agencies or organizations for which farmers are expected to pay a fee (or free) and it can be viewed as supplementary or alternative to public extension services.

Concept of privatization

Privatization of extension refers to services rendered in rural area and allied aspects of extension personnel working in private agencies or organization for which farmers are expected to pay a fee and it can be viewed as supplemented or alternative to public extension services (Sarvanan and Shivalinge, 1980).

Private extension services are primarily of two types

A. The first is the entirely private type which is directly involved in farming activities through consultants, agri-business, agricultural input industries etc.

B. The second type consists of farmer's organizations, NGO's etc., which remain largely dependent on government subsidies.

Privatization approach

- Share cropping system.
- Village extension system.
- Public extension through private delivery.
- Service for vouchers.

Agricultural extension programme using privatized service provision

- Agriclinic and agribusiness centres.
- Agribusiness initiative.

Favourable factors for privatization of agricultural extension in India

- Technological advancement in the areas of Communication Information Technology.
- Opening out of public research and educational institutes to part with the technologies.
- Changing – cropping trend and emergence of contact farming.
- Need to earn more from smaller holdings.
- Loss of credibility in the existing extension system.
- Inability of the public extension system to reach the large target.

Strengths of private extension system

- More demand – driven rather than supply – driven.
- Enhanced efficiency of health.
- High quality of service in terms of satisfying information needs of clientele, trained manpower, sustained finances and resource allocation.
- Provide forum information mix and choices available to farmers.
- Assure continuous supply and quality of agricultural products.
- Healthy competition among service provide will lead to better quality and lower costs for service.
- More effective because farmer can select an adviser who in the best able to help.

Problems in public extension system

- Basic thinking is supply driven rather than demand driven.
- Limitedly Focused on Farmers' Aspirations.
- Lack of location specific extension services.
- Extension services provided are general in nature.
- High cost, low impact of extension programmes.
- Poor motivation of staff working at grass root level.
- Curative nature of extension services.
- Inadequate technical qualifications of VLW.
- Incomplete extension services.
- Inadequate internal organization structure.

- Inefficiency of extension personnel.
- Inappropriateness or irrelevance of extension content.
- Dilution of impact.
- Deprivation of small farmers.
- Hamper the free flow of information.
- More face to face contacts (person oriented).
- Increased dependence of farmers and hence exploitation.
- Concentrate on area having favourable physical environment.
- No educate role.

■■■■

Chapter - 3 Extension Education

Extension - The word extension is derived from the latin word 'ex' - meaning 'out' and 'tensio' meaning 'stretching'. Stretching out is the meaning of extension. The word 'extension' came to be used originally in USA during 1914 which means "a branch of a university for students who cannot attend the university proper". In other words, the word extension signifies an out of school system of education.

Education: It is the production of desirable changes in knowledge (things known), attitude (things felt), and skills (things done), either in all (or) one or more of human behaviour.

Education is an integral part of extension. The basic concept of extension is that it is education. Extension means that type of education which is stretched out to the people in rural areas, beyond the limits of the educational institutions to which the formal type of education is normally confined.

Extension Education is an applied social science consisting of relevant content derived from physical, biological and social sciences and in its own process synthesised into a body of knowledge, concepts, principles and procedures oriented to provide non-credit out of school education largely for adults by Paul Leagans (1971).

Extension education is defined as an educational process to provide knowledge to the rural people about the improved practices in a convincing manner and help them to take decision within their specific local conditions by O.P. Dahama (1973).

"Extension Education is a science which deals with the creation, transmission and application of knowledge designed to bring about planned changes in the behaviour-complex of people, with a view to help them live better by learning the ways of improving their vocations, enterprises and institutions".

Extension education is an applied science consisting of content derived from research, accumulated field experiences and relevant principles drawn from the behavioural science synthesised with useful technology into a body of philosophy, principles, content and methods focussed on the problems of out of school education for adults and youths.

- Participation is purely voluntary.
- It is an informal education *i.e.* out of school and college.
- Extension worker started first practical work there after theoretical aspects.
- There is no fixed curriculum or course of study.
- Authority rests with the farmers.

- Teaching is mainly horizontal.
- Large and heterogenous audience.
- It is flexible and more practical oriented.
- It is a continuous education *i.e.* through out the life.
- Freedom to develop programme local need.
- Teaching by any kind of persons *i.e.* local leaders or lay leaders.
- It is Bi-dimensional learning.
- Farmers study the problems.
- It is based on the needs and expressed desires of the people.
- The word 'Learning by Doing' was coined by John Dewey in 1966.

Features of Extension Educations

The features of extension education are as:

i. Participation is purely voluntary.
ii. It is an informal education.
iii. Teacher or extension worker starts with practical first and thereafter theoretical aspects are to be touched *i.e.* first practical then theory.
iv. There is no fixed curriculum or course of study.
v. It is Bi-directional learning means teacher also learns from the farmers during the course of teaching to the farmers.
vi. It is flexible.
vii. Freedom to develop programmes locally.
viii. It is based on the needs and expressed desires of the people.
ix. It is continuous education *i.e.* throughout the life.

Objectives of Extension Education:

The main objectives of extension education are as:

1. Dissemination of useful and practical information.
2. Practical application of useful knowledge.
3. To improve all aspects of the life of the rural people.

Basic Philosophy of Extension Education

The basic philosophy of extension education are as:

i. Self help.
ii. People are the greatest resource.

iii. Programme of change.

iv. Cooperative effect.

v. Democratic nature means development of people by the people and for the people.

vi. Two channel of knowledge and experience.

  a. **Problems :** are brought to laboratory

  b. **Solutions:** are taken to farmers/villagers

vii. Programme is based on creating interest by seeing and doing

viii. There is voluntarily cooperative participation.

ix. Programme is based on attitude and values of the people with whom it works

Importance of Extension Education

The importance of extension education are as:

i. Extension uses democratic methods in educating the farmers.

ii. Extension helps in adoption of innovations

iii. Extension helps in studying and solving the rural problems

iv. Extension increases farm yields and improve the standard of living of farmers.

v. Extension makes good communities better and progressive

vi. Extension contributes to national development programmes.

vii. It develop, strengthens and organizes the groups, institutions and people to achieve their objectives.

viii. It develops leadership in local and professional situations.

ix. It motivate people to adopt innovations.

x. It builds confidence through action and conviction.

xi. It is a process of informing people leading to behavioural changes in the desired direction.

xii. It acquaints the planners, policy makers and administrators with local conditions and the latest technologies suited to it.

xiii. It gives direction and a package of educational practices for adoption and diffusion.

xiv. It improves the decision making ability of the people.

xv. It creates congenial learning situation and develops people own programme.

xvi. It suggests alternative fields for the needs and resources of the people.

CONCEPT OF THE EXTENSION EDUCATIONAL PROCESS

The concept of extension education processes are developed by Dr. J. Paul Leagans. These are:

1. The Education starts with study of present situation and identification of problems.

2. Based on the problems identified, solutions are found out and objectives are formulated.
3. In the third phase, plan of work is taught on how to attain the objectives, and then the entire process is evaluated and in the final stage reconsideration about the unattained objectives and unaddressed problems is done.
4. Extension education is an applied behavioural science to bring about desirable change usually through various strategies and programmes of change and by applying the latest scientific and technological innovations.

Scope of extension education

- Extension education is a science which deals with the creation, transmission and application of knowledge designed to bring about planned changes in the behaviour complex of people, with a view to help them live better by learning the ways of improving their vocations, enterprises and institutions.
- There is some logic in each of the above concept which may be defined as follows;

The different concepts may be defined as:

Extension service: An organization and/or a programme for agricultural development and rural welfare which employs the extension process as a means of programme implementation.

Extension process: The extension process is that of working with rural people through out of school education, along those lines of their current interests and needs which are closely related to gaining a livelihood, improving the physical level of living of rural families and fostering rural community welfare.

Extension Job: The job of extension in agriculture and home economics is to assist people engaged in farming and home making to utilise more fully their own resources and those available to them, in solving current problems and in meeting changing economic and social conditions.

- In other words, through the educational and services approach rural people are stimulated to make changes that result in more efficient production and marketing of farm products, conservation of natural resources, more comfortable homes, improved health and more satisfying family and community life.
- The modern concept of extension education is the means for achieving community development, includes several facets and subject- matter fields, of which agriculture is more important than the others. Hence "Agricultural extension" may be defined as special branch of extension education which deals with the economic and social aspects of people engaged in or associated with agriculture.

Extension appears to have unlimited scope in situations where there is need for creating awareness amongst the people and changing their behaviour by informing and educating them.

Kelsey and Hearne (1967) identified nine areas of programme emphasis, which indicate the scope of agricultural extension.

1. Efficiency in agricultural production.
2. Efficiency in marketing, distribution and utilization of farm produce.

3. Conservation, development and use of natural resources.
4. Management on the farm and in the home.
5. Family living.
6. Youth development.
7. Leadership development.
8. Community development and rural area development.
9. Public affairs.

Extension is an integral part of agricultural and rural development programmes in India. The progress in production which has been achieved in agriculture, horticulture, animal husbandry, veterinary, fishery, social forestry, sericulture etc., may be as proportional to the strength of extension service of the relevant government departments.

The following statements will further amplify the scope of extension.

1. Extension is fundamentally a system of out-of-school education for adults and youths alike. It is a system where people are motivated through a proper approach to help themselves by applying science in their daily lives, in farming, home making and community living.
2. Extension is education for all village people.
3. Extension is bringing about desirable changes in the knowledge, attitudes and skills of people.
4. Extension is helping people to help themselves.
5. Extension is working with men and women, boys and girls, to answer their felt needs and wants.
6. Extension is teaching through "learning by doing" and "seeing is believing".
7. Extension is working in harmony with the culture of the people.
8. Extension is a two-way channel; it brings scientific information to village people and it also takes the problems of the village people to the scientific institutes for solution.
9. Extension is working together (in groups) to expand the welfare and happiness of the people with their own families, their own villages, their own country and the world.
10. Extension is development of individuals in their day-to-day living, development of their leaders, their society and their world as a whole.

The need for extension arises out of the fact that the condition of the rural people in general, and the farm people in particular, has got to be improved. There is a gap between what is-the actual situation and what ought to be the desirable situation. This gap has to be narrowed down mainly by the application of science and technology in their behaviour. According to Supe (1987) the researchers neither have the time nor are they equipped for the job of persuading the villagers to adopt scientific methods and to ascertain from them the rural problems. Similarly it is difficult for all the farmers to visit the research stations and obtain first hand information. Thus there is need for an agency to interpret the findings of the research to the farmers and to carry the problems of the farmers to research of solution. This gap is filled by the extension agency.

PRINCIPLES OF EXTENSION EDUCATION

The principles of extension education are:

1. **Principle of interests and needs:** To be effective, extension work must begin with the interest and needs of the people. Extension agency mould the needs and interests of the people into realistic needs. The need and interests differ from individual to individual from village to village, block to block and state to state, therefore, there cannot be one programme for all people.
2. **Grass root principle of organization :**To be effective and real, it has to be synthesis of democracy obtained at the level of the family and more particularly at the village level. The programme should fit in with the local conditions. The aim of organizing the local group is to demonstrate the value of the new practice or programmes so that more and more people would participate.
3. **Principle of cultural differences:** Cultural differences exist between extension worker and farmer. To be effective the approach and procedure must be suited to the culture of the people who are taught.
4. **Principles of cultural change:** extension education starts with what the learner knows, has and think. With this in mind and with an attitude of respect towards clients, the extension professionals must seek to discover and understood the limitations, taboo and cultural values related to each phase of programme so that an acceptable approach could be selected in the locality.
5. **Principle of cooperation and participation:** The participation of the people is of fundamental importance for the success of any educational endeavour. Ultimately without the cooperation of people the work cannot be successful and desire results cannot be achieved.
6. **Principles of applied science and democratic approach:** Extension education is based upon democratic principle. It is based on discussions and suggestions. Discussions are held with the people on actual field conditions, so that they participate in work. Extension professionals provide practice look to the scientific inventions so that farmers can easily adopt them on their farm.
7. **Principle of learning by doing:** Learning by doing involves use of maximum number of senses, hence it is very effective in changing behaviour. According to this principles encouraged to learn by doing the work themselves and by participating to it, when a person does a work, he gains practical knowledge and experiences the difficulties. Extension professionals are able to understood the problems and provide guidance to the farmers and thus, they are able to receive proper information/feedback.
8. **Adaptability principle in the use of extension teaching methods:** The use of teaching methods must have flexibility to be adopted to the members of a community who differ in age, education, economic status, sex and proneness to change etc. In extension education, two or more methods should be applied according to the principle of adaptability.

9. **Principles of trained specialists:** It is very difficult that extension personnel should be knowledgeable about all problems. Therefore, it is necessary that specialists should impart training to the farmers from time to time.
10. **Principle of leadership:** The involvement of leaders in extension programmes is the one single factor that determines the success or failure of those programmes. Extension work is based on the full utilization of local leadership. The selection and training of local leaders to enable them to help in carrying out extension work is essential to the success and they should be used to put across a new idea so that it is accepted with the least resistance.
11. **Whole family principle:** The family is the unit of any society. Creates mutual understanding, aids in money management, balances farm and family needs, educates the younger members, assures family service to the community and society. Extension work is therefore, for the whole family, *i.e.* for male, female and children.
12. **Principle of evaluation:** Extension is based upon the methods of science and it needs constant evaluation. The effectiveness of the work is measured in terms of the changes brought about in the knowledge, skill, attitude and adoption behaviour of the people, not merely in terms of achievement of physical targets.
13. **Principle of satisfaction:** Unless the people are satisfied with the end product of any programme, it is not going to be able to run. Satisfaction is the key success in extension work.

It take into account that the various problems emerging at the social level due to implementation of developmental programmes.

The discipline of extension education is to generate knowledge. Extension lays emphasis on "How to teach" instead of on "What to teach" so that the people can be encouraged to adopt new research techniques easily.

EXTENSION EDUCATION PROCESS

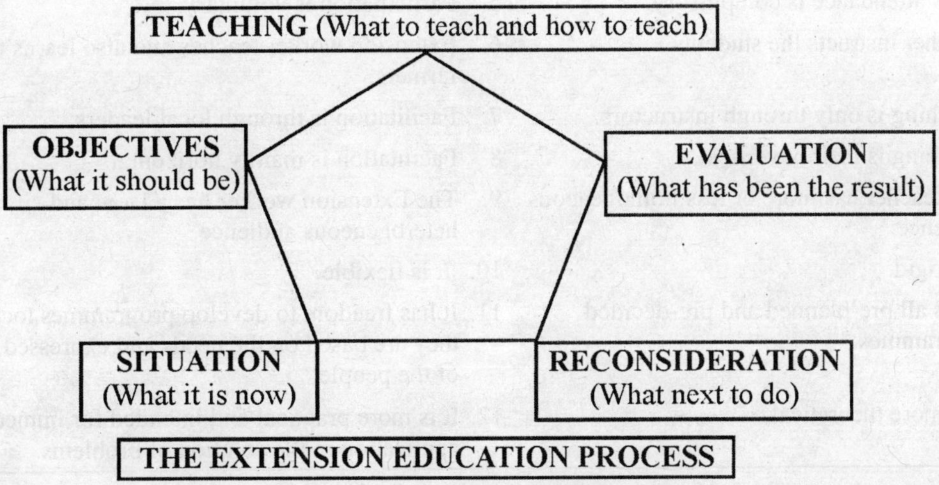

THE EXTENSION EDUCATION PROCESS

The effective extension education programme involves five essential and interrelated steps these are:

First step: The first step consists of collection of facts and analysis of situation. Facts about the people and their enterprises, the economic, social, cultural, physical and technological environment in which they live and work.

Second step: The next step is deciding on realistic objectives which may be accomplished by the community. The objectives should be specific and clearly stated and on completion should bring satisfaction to the community, objectives should state the behavioural changes in people as well as economic and social desire.

Third step: The third step is teaching, which involves choosing what should be taught (the content) and how the people should be taught (the methods and aids to be used). It requires selecting research findings of economic and practical importance relevant to the community and selecting and combination of appropriate teaching methods and aids.

Fourth step: The fourth step is evaluating the teaching *i.e.* determine the extent to which the objectives have been reached.

Fifth step: The fifth step is re-consideration of the entire extension educational programme on the light of the results of evaluation.

Difference between Formal Education and Extension Education

	Formal Education		Extension Education
1.	The teacher starts with theory and works up to practical.	1.	Extension worker starts with practical and may take up theory later on.
2.	Students study subjects.	2.	Farmers study problems.
3.	Students must adapt themselves to the fixed curriculum offered.	3.	It has no fixed curriculum or course of study and the farmers help to formulate the curriculum.
4.	Authority rests with the teacher.	4.	Authority rests with the farmers.
5.	Class attendance is compulsory.	5.	Participation is voluntary.
6.	Teacher instructs the students.	6.	Extension worker teaches and also learns from the farmers.
7.	Teaching is only through instructors.	7.	Facilitation is through local leaders.
8.	Teaching is mainly vertical.	8.	Facilitation is mainly horizontal.
9.	The teacher has more or less homogeneous audience.	9.	The Extension worker has a large and heterogeneous audience.
10.	It is rigid.	10.	It is flexible.
11.	It has all pre-planned and pre-decided programmes.	11.	It has freedom to develop programmes locally and they are based on the needs and expressed desires of the people.
12.	It is more theoretical.	12.	It is more practical and intended for immediate application in the solution of problems.

Role of Extension Education

Extension education centres around three units:

(a) Farmers or rural people

(b) Innovations or inventions

(c) Extension workers

The technological development of education can be related to various fields of extension like agriculture and home Science.

There are however, certain broad role of extension education which can made it. These are:

1. Extension work can be most effective only when it is entirely educational in nature and free from regulator and other government responsibilities.
2. As development proceeds, more specialization at the field level may be feasible and desirable.
3. The successful work of the extension officer at the field level depends upon their being supported by specialists at the next higher level of administration.
4. Efforts should be made to keep extension and research closely related to each other.
5. Extension work must be directly towards the family as a whole men, women and children work with youth and farm women is important.
6. In the interest of providing maximum effective service within the limited resources, then is an urgent need for effective co-ordination of all the agencies serving rural people.

■■■■

Chapter - 4 Extension Methods

It is important that the educational purpose of extension work, and to ensure that the method selected is used to promote the farmers' better understanding of the technology involved. Both individual and group extension methods involve the agent in a face-to-face relationship with the farmer, and this relationship should be one of mutual confidence and respect. The agent, therefore, must think carefully about his use of individual or group extension methods and ensure that his relationship with the farmer is sensitively developed.

Types of Extension Methods

These are broadly categorized into three: Individual, Groups and Mass methods.

A. Individual Methods of Extension

Individual or face-to-face methods are probably the most universally used extension methods in both developed and developing countries. The extension agent meets the farmer at home or on the farm and discusses issues of mutual interest, giving the farmer both information and advice. The atmosphere of the meeting is usually informal and relaxed, and the farmer is able to benefit from the agent's individual attention. Individual meetings are probably the most important aspect of all extension work and invaluable for building confidence between the agent and the farmer.

The individual contact between the extension agent and the farmer can take a number of forms such as farm and home visits, office calls, personal letters, phone calls, e-mails and its related means like messenger/facebook.

1. Farm visits

Farm visits are the most common form of personal contact between the agent and the farmer and often constitute over 50 percent of the agent's extension activities. It is important to be clear about the purpose of such visits and to plan them carefully.

- familiarize the extension agent with the farmer and his family.
- enable him to give specific advice or information to the farmer.
- build up the agent's knowledge of the area and of the kinds of problems which farmers face.
- permit him to explain a new recommended practice or follow up and observe results to date.

- arouse general interest among the farmers and stimulate their involvement in extension activities.

The informal visits may have no specific purpose but are a useful way of maintaining contact and gradually building links with farmers. Farm visits will be part of the agent's general plan of work and will be programmed into his monthly schedule of activities.

It is important to be clear about the purpose of the visit. Before a visit, the extension agent should review the file on the farm to be visited and consult any other information available on the farmer.

The visit should be planned to fit in with other local extension activities. For example, if a demonstration or meeting is planned in the morning, then it may be possible to programme a number of individual farm visits in the afternoon.

The following points should be taken into consideration for planning of visit:

- Make an appointment if possible.
- Decide the purpose of the visit.
- Review previous records and information.
- Prepare specialist subject matter that might be required.
- Schedule the visit into the overall work plan.

The visit may cover a whole range of activities. The farmer may need further explanation or information about a particular new practice and it may be necessary to demonstrate this. If further technical information is required, this can be noted and at a later date, an appropriate information bulletin and relevant material should be provided. The agent may also brief the farmer generally on government agricultural policy or describe the range of extension activities in the area and encourage the farmer to get involved. The extension agent should always be alert during farm visits for ways to promote the involvement of other family members, including spouses and young people, in other local extension activities such as a youth club or a food preparation demonstration.

If possible, it is useful to set up a system to keep a record of the content of farm visits; this could be simply a notebook in which the essential details of each visit (date, purpose, problems, decisions, etc.) are recorded. Such a farm record system is very useful in helping the agent to keep a more detailed account of each visit and it is invaluable when a new agent is taking over. The records are also very useful in evaluating farmers' progress.

The following points should be taken into consideration for making the visit:

- Be punctual for the visit.
- Greet the farmer and this family.
- Praise the farmer's work.
- Encourage the farmer to explain and discuss any problems.
- Provide any technical or other information required.
- Record the details of the visit.
- Plan with the farmer - the time and purpose of the next meeting.

2. Office Calls

The extension agent visits the farmer, so he can expect that from time to time the farmer will visit him at his office. Such a visit is often a reflection of the interest which the agent may have aroused among the local farmers. The more confidence local farmers have in the extension agent, the more likely they are to visit him. Such office visits are less time consuming for the extension worker and offer some of the advantages of a farm visit.

As with farm visits, office visits similarly have to be prepared. Although the agent may not know when a farmer is likely to drop in, he can at least arrange the extension office in such a way that the visitor feels at ease and can understand the activities of the office. The arrangement could include:

- ensuring that access to the extension office is adequately posted and the agent's name displayed;
- having a notice-board clearly displayed upon which useful, up-to-date information can be pinned;
- having one or two chairs where visitors can wait for appointments;
- displaying any bulletins, circulars or other written extension literature that the visitor can read.

3. Letters

Occasionally, the extension agent will correspond with a farmer by letter. Letters can be a follow-up inquiry resulting from an agent's farm visit, or sent because a farmer is unable to make a personal office visit. Drafting and replying to letters are very important skills for the extension worker and he should give every thought to them. In writing a letter to a farmer, the extension agent should try to put himself in the farmer's shoes. The letter should be in the local language, preferably not on impressively headed writing-paper, and should always contain some personal greeting to the farmer. Often, farmers will show such letters to their neighbours and thus it is important to create a favourable impression. The following points are important:

- letters should be clear and concise, so as not to confuse the reader.
- the information in the letter should be complete and relevant to the issues raised.
- where possible, letters should be answered promptly. If time is needed to collect information for the reply, a short letter of acknowledgement should be sent.
- a copy of the letter must always be made and entered in the office file.

4. Telephone Calls

Telephone calls and office visits serve a very similar purpose. It is improbable that the extension worker will deal with many of the farmers in his area by telephone. If the telephone is used, however, it will not be for long discussion but for passing on specific advice or information. Whatever the reason, it is important for the agent to speak clearly, to note the main points discussed and to enter them on the farmer's record.

5. Informal contacts

Informal contacts will occur continuously during the agent's stay in a particular area. Market days, holiday celebrations or religious events will bring him into contact with the farmers with whom he is working will inevitably talk about their problems. By attending such events, the agent can become well acquainted with the area where he works and with the farmers and their problems, and he will be able to pass on ideas and information on an informal basis.

6. Extension talk

Extension talk differ from a normal conversation, an enquiry or an interrogation. It is always the client and his/her way of perceiving a problem that is the focus and starting point of the extension worker. These are formal and informal extension talks:

1. Formal talks usually take place on the form or in advisor's office- the aim of such talks is to isolate concrete problems and work out ways of solving them .
2. There are informal contacts that can occur on numerous occasions and the extension worker may well find this the easiest way of getting to understand the wishes and difficulties of client groups and of establishing contact with them.

- The course and the function of extension talks be determined by the objective of clearly identifying existing problems and their causes and trying to arrive at possible solution together. A discussion will only be productive if there is a common way of looking at a problem.
- The adviser must not be tempted to impose a solution on the farmer even though he might be in a position to do so on account of his superior knowledge and power of expression.
- Understanding problem becomes easier when the person to be advised is expected as on individual.
- It is not uncommon to find prejudice, a desire to impress and a disapproving attitude on the part of the extension agent, which may result to a defensive attitude on the part of the client and make extension advise an impossible.

7. e- mail

- Meeting personally is absolutely necessary at the beginning of an adviser/client relationship.
- Later on, if a positive and trusting relationship exists, not every activity must involve personal contact.
- Where as possible, shorter message and standard information can also be transmitted impersonally.
- For example SMS, fax and e-mail call are immediate means of contact that allow for a direct exchange of information and intermediate response to a message, meaning that misunderstanding can be better prevented compared with one-way communication.

- The diffusion of mobile telephone networks even to remote areas makes things much easier. But the more critical places is in complex problem which solving that calls for physical presence and direct personal contact.

8. Market place or at religious places

- Being present at the market or near church or mosque or temple may make easier for the clients to approach the adviser.
- The hurdle the clients have to overcome may be lower compared to calling by telephone or going to the office, so that after a few words of greeting and small talk, the client can indicate that he/she has problems and need support. Then an appointment for a meeting can be made.
- Beginning to talk about the problem immediately is generally not appropriate, the solution is too public, not confidential enough and too much noise and disturbance limits the efficacy of interaction.

B. Group methods of extension

The extension agent should consider the use of the group approach in his work with farmers. The use of groups in extension has become more common over the past decade and indeed a number of new ideas have emerged about how groups may be used most effectively. For example, the widespread Small Farmer Development Programme (SFDP) in Southeast Asia was based upon group methods and it has produced two manuals which detail the approach of group extension work. Furthermore, in Latin America, work with extension groups in Brazil and Colombia has shown the usefulness to extension of the formation of extension groups and how these groups can support extension activity. It has been seen that individual extension methods can be costly in both terms of time and scarce extension resources and that they reach only a limited number of people.

The group method offers the possibility of greater extension coverage, and is therefore more cost-effective. Using the group method, the extension worker can reach more farmers and in this way make contact with many more farmers who have no previous contact with extension activities.

Extension groups offer a more reflective learning environment in which the farmer can listen, discuss and decide upon his involvement in the extension activity. The support of the group helps an individual farmer to make decisions and determine a course of action.

- Several members of the client (farmer) group who are linked by formal and informal ties are addressed at the same time.
- It is the most important method for advising and promoting the interest of a large number of clients (farmers).
- The more it is supplemented by individual and mass extension methods, the greater are its chances of success.
- However, group extension method is not always the best way of communicating particularly difficult and complex subject matters.

- Group extension makes increased demands on the methodological and organizational skills of the adviser.
- Unless groupings of farmers already exist, it is difficult to form such groups and keep them functioning.
- The adviser cannot determine the composition of group, his/her task is, among other things, to improve cohesion of groups, the group climate and how they function by means of continuous training and counselling.
- Group methods require good planning in selecting and inviting the participants in determining and preparing programme.
- The use of teaching and extension aids makes group methods more attractive and more effective.
- The presented of specialists and representatives from complementary organization is often necessary and calls for advance planning and arrangements.
- Compared to individual extension work, there is a greater chance of conflicts between clients (farmer) and advisers, but this must be viewed positively; in groups clients can emphasize their interest to a greater effect.

Types of Group Extension Methods

The examples of group method of extension are: Demonstration, meetings, group/panel discussions field tour/trips, field day, farmers field schools, farmer' training centres, seminars/ workshops etc.

1. Group Meetings

Calling the members of a group or the inhabitants of a local community together for a meeting is the commonest group extension method. Although there may be an air of informality about such meetings, they will nevertheless need to be carefully thought out and planned. The group or community meeting is a useful educational forum where the agent and farmers can come together and ideas can be openly discussed and analysed. The agent will probably have information about a new government policy or agricultural idea or practice. He will want to introduce this new information, to seek the opinions of community members and gain their support for extension activities. Indeed, there are a whole range of purposes for such community or group meetings:

i. **Information meetings:** The agent calls the group or community together to communicate a specific piece of new information which he feels will benefit them and upon which he seeks their advice.

ii. **Planning meetings:** The main purpose is to review a particular problem, suggest a number of solutions and decide upon a course of action.

iii. **Special interest meetings:** Topics of specific interest to a particular group of people (e.g., horticulture, bee keeping, or dairy farming) are presented and discussed in detail at a level relevant to those participating.

iv. **General community meetings:** Men, women and young people of a community are invited to attend to discuss issues of general community interest. It is important to hold such general meetings occasionally so as to avoid any community group feeling that it is excluded from extension activities.

The basic purpose of the meeting should be agreed and to determine this the agent should consult community or group leaders. The agent and community leaders consider the content and the best approach to the meeting.

Depending upon the nature and purpose of the meeting, the agent must decide the most appropriate form for the meeting and how it can best be conducted. Smaller meetings are more likely to meet the specific needs of those who attend.

There are two important decisions to make regarding the time and location. A date and time for the meeting must be decided and announced. The time should be convenient to all concerned and should avoid clashes with other events or activities. The meeting-place should be well-known, easy to get and appropriate for the meeting. The meeting-place should also be comfortable and have the facilities necessary for the meeting. An extension agent would never hold a meeting at midday, on a very hot day, in the open sun. Such a meeting could be disastrous, as well as cause considerable discomfort.

After the above two issues have been considered, it may be useful for the agent to draw up a list of other arrangements to be made in preparation for the meeting. Such a check-list could include:

- Publicity for the meeting
- Seating arrangements
- Audio-visual equipment and material, or other educational aids
- Agenda, and order of events
- Guest speakers or other specialists who will contribute to the meeting
- Chairman to take charge of the meeting, who should be elected by the community
- Refreshments for speakers and, if necessary, for other participants. also.

The agent should try to vary the agenda of the meeting: for example, a short talk, accompanied by visual aids followed by comments and questions.

Variety of content, as well as a chance for the farmers to participate, will be important. In addition, the meeting must not go on too long. One-and-a-half hours are probably sufficient for a group or community meeting. It is better to have a highly productive, short meeting than one which rambles on and loses effect.

He should encourage the community to appoint a chairman and should allow the chairman to conduct business. The agent's role should basically be to inform and support, and he should not dominate the meeting.

As a guide to the proceedings of the meeting, the agent should keep the following points in mind. He should start the meeting on time. Then he should welcome community members and special guests, explain the purpose of the meeting and the programme to be followed and begin the programme. Later, the agent should encourage questions and discussion and be prepared to summarize the main points and note important decisions. The meeting should be closed with thanks to all concerned.

2. Demonstrations

Farmers like to see how a new idea works, and also what effect it can have on increasing their crop production. Both purposes can be achieved by means of a farm demonstration. A good, practical demonstration is an invaluable method in extension work. The demonstration is a particularly powerful method to use with farmers who do not read easily. A demonstration will give such farmers the opportunity to observe, at first hand, the differences between a recommended new crop practice and traditional practices. The strength of the demonstration should lie in its simplicity and its ability to present the farmers with concrete results.

There are mainly two types of demonstration used by extension agents - method demonstration and result demonstration.

(i) Method demonstration

Method demonstrations basically show farmers how to do something. In the method demonstration, the farmer is shown step by step how, for example, to plant seeds in line, to use a mechanical duster to control insects, or to topping of tobacco. The agent will probably be dealing with farmers who have already accepted the particular practice being demonstrated, but who want to know and how to do it themselves.

The main advantage of the method demonstration is that the extension agent can explain simple farming skills to a large number of people, thus increasing the impact of his extension work. Moreover, as farmers are able to participate, there is a greater chance that they will benefit from the demonstration than if they were passively hearing it in a lecture.

The main limitation of a method demonstration is that, if there are too many farmers present, only a few get a chance to see, hear and do. The agent must be conscious that the demonstration is a learning experience and prepare the event accordingly. It is also vital that the demonstration be well thought out and competently conducted.

(ii) Result demonstration

The main purpose of a result demonstration is to show local farmers that a particular new recommendation is practicable under local conditions. Comparison is the important element in a result demonstration: comparison between compost and no compost, between poor seed and selected seed, or between use of fertilizer and no fertilizer. "Seeing is believing" is an age-old expression, but one appropriate to a result demonstration. Until a farmer has actually seen the results of, for example, the application of a fertilizer, he will not be convinced by the agent's recommendation. By showing tangible results of a new practice recommended by the extension service, the agent can help to create confidence among the farmers and can greatly encourage them to try the practice themselves.

A result demonstration is an ideal way to present to farmers a comparison between traditional and new practices. It can also help to establish confidence in more scientific farming methods and increase the farmers' confidence in ideas originating from research stations. It shows proof of the value of a new practice. A result demonstration is also a useful tool that an agent can use to establish confidence among farmers in a new area.

Its major limitation is that it takes a long time to mature and is thus a costly use of extension resources. If, in the end, for whatever reason, the new practice should fail, it could have disastrous consequences. Often such failures (for example, because of lack of rain) are outside the control of the agent.

Both method and result demonstrations are extension activities that require a lot of thought, careful planning and efficient execution. Although the two demonstrations differ somewhat in their purposes, they share a lot of common points and, in terms of their preparation and execution, they can be considered together.

3. Field Days

Field days are usually opportunities to hold method or result demonstrations on a slightly larger scale, and are usually run in a more informal and less highly structured manner. The purpose is often to introduce a new idea and a new crop and to stimulate the interest of as many farmers as possible. Experimental stations or other government centres may be used for field days, but it is more usual and profitable for them to be held on the land of a local farmer.

Field days can vary in size from a small group to annual events attracting hundreds of farmers. The extension agent's role on the field day is to support the farmer on whose land it is being held, to offer general guidance to ensure that things run smoothly and to be available to answer questions and queries.

4. Tours

Farmers like to visit farms in other districts to see how they work, what they grow and what kinds of problems the farmers there are facing. A tour is a series of field demonstrations on different farms, or at different centres, and can often attract a lot of interest from local farmers. The tour should give local farmers a chance to see how other farmers cultivate their land, and to exchange ideas and experiences with them. It is important, therefore, that the area to be visited be in some way similar agriculturally to that of the visiting farmers.

As with all other forms of extension, tours have to be well thought out, planned, prepared and conducted. The five stages of determining the objective, planning the content, preparing the arrangements, conducting the tour and arranging for appropriate follow-up will be a guide to the extension agent.

A field tour is an ideal method of involving farmers and of stimulating genuine interest in extension activities. It is also very useful in bringing farmers together to discuss common problems and to gain useful experience of other areas.

5. General Meetings

One of the oldest and most important methods of extension teaching. Meetings are called by extension worker for training about the production technology of the next crops as well as discussing the various problems being faced by farmers/clients. Adequate time an encouragement should be provided to the farmers/clients to discuss their difficulties.

- General meeting include all kinds of meeting held by the extension worker except demonstration meetings.

  During the meeting provision should be made for use of

  – Models

  – Charts

  – Specimen

  – Pictures etc to illustrate points
- Towards the end of the meeting arrangement should be made for questions and answer which would help clarify specific ideas.
- Calling members of a group or the inhabitants of a local community together is the most commonest group extension method despite level of formality meeting are to be well thought out and planned.
- There are a range of purposes for such meeting including:

  – Informing (information meeting)

  – Planning (planning meeting)

  – General (community meeting)

6. Group discussion

The process by which or more persons pool their knowledge and feeling through mutual agreement and clarify the issues under consideration. Through discussion various points of view and the interests of the participants are aroused to do something towards solving their problems.

Key issues include

a. A group

b. Working together (sharing)

c. A topic of interest

- Existing practices should not be criticized instead new ideas should be placed before the group for discussion.
- The extension worker should avoid giving the impression that his/her solution is the best.
- Avoid imposing the pre-conceived idea on the people.
- In the initial step it may be unsuccessful, therefore, the extension worker should not get frustrated.
- Before going for a meeting, a well prepared concept and a clear idea may be born in mind.

C. Mass Method of extension

Mass methods are those channels of communication which convey information to a large number of people at a time. They include media which convey information through print (posters, newspapers, leaflets), by sound (radio) and moving pictures (TV and films – on tapes or VCD/DVD and mobile phones).

There are different functions of mass media in extension education:

1. Spreading information awareness and creating interest in a community especially to the innovators/venturesome.
2. Giving timely warning about a possible event like pest and diseases outbreaks and urgent advice on what action to take.
3. Multiplying the impact of extension activities. For example, a demonstration will only be attended by a small number of farmers but results will reach many more, if they are reported in newspapers and radio or television on TV or multiplied in VCD/DVDs and distributed to various groups of client with similar interest.
4. Sharing experiences with other individuals and communities. The success of a village in establishing (example – a tree plantation may simulate other villages to do the same if it is broadcast over a radio. Farmers are also interested in hearing the problems of other farmers/clients and how they have overcome them.
5. Answering questions and sharing on problems common to a large number of farmers/clients.
6. Repeating information and advice. Information heard at a meeting or passed on by an extension agent can soon be forgotten. It will be remembered more easily if it is repeated by mass media.
7. Farmers/clients can be brought into contact with successful farmers/clients from other areas, respected leaders and agricultural experts through mass media.
8. In mass media communication requires specialist professional skills. Extension agents can contribute to the successful use of mass media by providing material to media producers in the form of newspaper stories, photographs, recorded interviews with farmers items of information about extension activities or ideas for new extension works e.g. by distributing posters and leaflets or by encouraging farmers to listen to farm broadcast on radio and TV.

The principles of media use are:

For effective extension work through mass media, the farmers/clients:

- Have access to the medium
- Be exposed to the message, they may have radios/TV, but do they listen to broadcasts related to farming or watch agricultural programs on TV?
- Pay attention to the message information. It must be attractively presented and relevant to farmers'/clients' interest.
- Understand the message.

Types of Mass Media Materials/Methods

The mass media broadly classified into two groups as:

1. Print media – posters, newspapers, magazines, leaflets, circular letters.
2. Audio-visual media – radio/TV programs for a specific subject matter, mobile phone.

1. Print media

The term print media is used to cover communication techniques that rely principally on combination of printed words and pictures. They are our oldest formal combinations. To use them effectively, the educational levels and literacy rates of the audience must be considered.

Posters

- They are useful in publicizing forthcoming events and for repeating messages that farmers/clients receive through other media.
- Prepared as part of (agricultural) campaigns and issuing warnings or instructions against outbreaks or losses.
- They vary in size and may also be supported with appropriate photographs.
- They should be displayed in prominent places where a lot of people pass by regularly.
- The most effective posters carry a simple message, catch people's attention and are easy to interpret.
- Posters are put on walls of buildings, fences, trees, bulletin boards, trucks, automobiles etc.

Newspapers

- Not widely available in rural areas. However, local dealers often read newspapers and a regular column on agricultural topics is useful to create awareness of new ideas and to inform people of what other groups or communities are doing.
- Now-a-days many newspapers publish a special weekly page on agriculture.

Magazines

- These are published on monthly or quarterly basis by the agricultural departments to convey new information or seasonal calendars.
- Farmers' organizations, association and chambers also publish periodic magazines preferably in widely used local language.

Leaflets

- They can summarize the main points of a talk or demonstration or provide detailed information that would not be remembered simply by hearing it, such as fertilizer application rates or names of crop varieties, new/improved cultural practices of crops, plant protection measures, introduction of new farm implements etc.

Circular letters

- They are used to publicize local extension activities to give timely information on local farm/community problems and to summarize results of demonstrations so that the farmers/clients who can not attend them may still benefit.

2. Audio-visual aids

Radio – radio is a particularly useful mass medium for extension. They may be electrically or battery operated.

- Information can reach households directly and instantly throughout an area.
- Urgent news and warning can be communicated more quickly on radio than print media.
- Radio is the best source of spreading new ideas and activities of extension to a large number of people at a time.
- It can enable one distinct or group to share its experience with others.

Television/VCD/DVD – combined vision (picture) with sound:

- TV broadcasts are limited to far-flung /remote areas.
- Moreover majority of poor farmers/clients can not afford buying television sets.
- Now-a-days TV centres are telecasting films or agriculture/nutrition regularly to teach farmers/community members.
- Video/films can also be made locally to show to the farmers/community members.

Mobile phones – They found that offering mobile extension services proved to be a cost effective mechanism to nudge farmers towards more productive agricultural practices.

- Mobile phone usage in third world countries is playing a vital role for the enhancement of farmers business towards agriculture.
- Mobile phone saved energy and time of farmers and ultimately improve their income.
- Smartphones and their applications are innovations bringing good solution for agriculture development in other to help farmers to have access to relevant informations.
- It participate in the amelioration of agriculture extension work and advisory services.
- Samartphones and its applications has come with great innovations. The application have been developed to help farmers reduce stress, acquire relevant informations on good agricultural practices, weather, quality input, market tendency etc.
- Through social media, web sites and other applications, farmers can improve their skills, share experiences and even sale their products online using their smartphones.

■■■■

Chapter - 5 **Teaching-Learning Process**

The teaching-learning process is essential to have a grasp of the meaning of extension.

Teaching: Is the process of arranging situation in which the important things to be learned are called to the attention of the learners, their interest developed, desire aroused and action promoted.

Principles of Teaching

i. The learner should subscribe to and understand the purpose of the course.
ii. The student should have desire to learn.
iii. The teacher should keep friendly and informal relationship with learner.
iv. The physical condition should be favourable and appropriate to the learning situation.
v. The teacher should involve the learners so that they participate and accept responsibility for the learning process.
vi. The teacher should make use of the learner's experiences.
vii. The teacher should prepare ahead of the class, keep his teaching aids handy and should be enthusiastic about teaching it.
viii. The method of instruction should be varied and appropriate.
ix. The teacher should update his notes with the availability of new knowledge on the topic of subject.

Leaning

It is the process by which an individual, through one's own efforts and abilities change the behaviour.

Learning is the process by which a person becomes changed in his behaviour through self-activity.

"Learning is the process of progressive behaviour adaptation".

1. Learning is something that takes place within the learner.
2. It takes place within the individual when he feels a need, strives for fulfilling it, and experiences satisfaction with the fruits of his labour.
3. Learning is the goal of teaching. We must however avoid any artificial separation, since teaching and learning are really one process; they are so as to say, two sides of the same coin.

Process: Means a course of procedures, some things that occurs in a series of action or events conducting to the desired end.

Learning Experience

It is mental and/or physical reaction one makes through seeing, hearing or doing things to be learned, through which one gains meanings and understandings of the material to be learned. Learning is an active process on the part of the learner. Hence, learning experience is not attained by mere physical presence in a learning situation, it is what the participant does (*i.e.* his reaction) in the learning situation. He must give undivided attention to the instructor and deep-through for getting the facts, understanding their meaning, and to see their application to his needs and problems. Effective learning experiences therefore, can be best gained in effective learning situations provided a skilful instructor who knows what he wants, who has the material to accomplish his goals and the skills to use them effectively.

Learning Situation

Learning is a coordination of environment in which all the elements necessary for promoting learning are, (1) Instructor (2) Learner (3) Subject matter (4) Teaching materials and equipments and (5) Physical facilities.

The nature of each of these elements, their relationship to each other, their role in the educational process must be thoroughly understood by the instructor and skill developed by him in handling them. Effective learning situations are created through the skillful use of appropriate teaching methods and techniques.

EXTENSION LEARNING PROCESS

Agricultural extension is an educational process for bringing desirable changes among the agricultural farmers, which involves both learning and teaching and needs some tools or methods commonly known as extension-teaching methods. It is, therefore, necessary to understand what is meant by learning, teaching and extension methods.

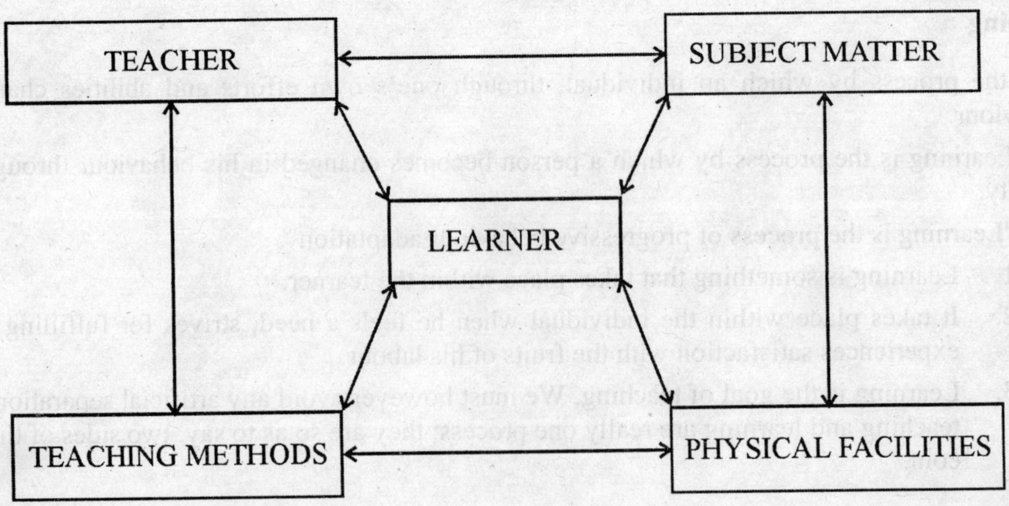

Any change of behaviour which takes place as a result of experience may be called learning. 'Learning' is the process by which an individual, through his own activity, attains a change in his behaviour (Leagans, 1961). It is an active process on the part of the learner. The essential role of an extension worker is to create effective 'learning situations'. An effective learning situation requires the following essential elements (Leagans, 1961):

Elements of Extension Learning Process

Teacher (Instructor)

A successful teacher is the one who takes into account the following important considerations:

- Understands the nature of the learner and is in a position to evaluate his motives and ability.
- Reacts appropriately to the feelings, emotions and attitudes of the learners.
- Is conscious of the fact that the learner responds to the whole gamut of learning situation.
- Be democratic in the leadership.
- Is professionally alert.
- Is well-composed even in the adverse conditions.
- Has the ability to maintain better human relations.
- Always makes efforts for his professional growth.
- Selects those experiences which suit to the abilities and needs of the learners.
- Should be close to masses and belong to the same groups for making dialogues easier.
- Be suited in the use of teaching materials and equipments.
- Be prepared, be prompt, be friendly and be counteous.

Learners

They should have:

- Clear objectives regarding the use of the subject matter.
- Interest in the subject matter.
- Need for the subject matter.
- Ability to understand and opportunity to apply the subject matter.
- Physical and mental preparation.

Subject Matter (contents)

The subject matter should fulfil the following conditions:

- Should be in the line with the needs and interests of the learners.

- Should be valid, authentic, factual and applicable.
- Should be well-organized.
- Should be at a level where it could be understood by the learners.
- Should be timely and appropriate.

Teaching Methods

Subject matter cannot effectively be transferred without the help of suitable teaching methods and aids. Proper selection and skilful handling of teaching methods and aids facilitate creation of desirable learning situation. Teaching methods and aids should be:

- Readily available.
- In working conditions.
- Suitable to the subject-matter.
- Diversified, flexible and suited to the needs of the clients.
- Challenges, satisfying and significant to the learners.

The following points need to be kept in mind while using extension methods for appropriate involvement of different sense organs:

30% What We See

↓

50% What We See and Hear

↓

70% What We Say

↓

90% What We Say and Do

↓

10% What We Read

↓

20% What We Hear

Physical Facilities

Light, ventilation, chair and table arrangements, outdoor facilities, gardening and good land escaping should be a satisfying and comfortable, if possible. The good teacher will make all these arrangements prior to the meetings.

The teacher should skilfully manipulate the elements of the learning situations and provide satisfactory learning experiences for the stakeholders. The main aim of the teacher is to bring about a change in the behaviour of the learners with the help of a judicious combination and use of different elements.

Criteria for effective extension Teaching

1. Extension teaching requires specific and clearly defined teaching objectives: Teaching must be clearly conceived and specifically defined. There must be clear-cut conception about the end product, *i.e.* what is going to be achieved.

In deciding on the objectives of teaching, the following aspects are to be considered:

i. People to be taught

ii. Behavioural changes to be developed in people

iii. Content or subject matter to bring the desired change in behaviour

iv. The real life situation in which the action is going to take place

2. Extension teaching requires a suitable learning situation: The learning situation consists of the following five interrelated elements - Instructor (scientist/extension agent), Learners (farmers, farm women, rural youths etc.), Subject matter (e.g. Horticulture, Dairy, Agriculture, Farm forestry or Fishery), Teaching materials (seeds, plant, audio-visual aids etc.) and physical facilities (suitable land, facilities for conducting demonstrations, training etc.). The learning situation of the extension teaching should be good.

3. Extension teaching requires effective communication: Communication involves the transfer of knowledge from a source to one or more receivers. Extension teaching to be effective, must take into consideration proper functioning of the elements of communication-communicator, message, channel, treatments and presentation, audience and audience response. Effective communication requires good feedback information.

4. Extension teaching requires both content and method: Content means what to teach, the subject matter, method means, how to teach, the delivery system. The content should be relevant to the audience. Appropriate teaching methods and aids are to be selected and combined according to needs of the situation.

5. Extension teaching must be looked upon as an intentional process: Extension teaching should be properly planned and designed on the basis of relevant data on the situation and available research findings. There is no scope for trial and error or haphazard thinking and action in extension teaching. Motivating people to learn is essential.

6. Extension teaching must result in effective learning: The meaning of what is taught should be understood and internalized by the learners. This usually requires a combination of teaching methods and aids relevant to a particular situation. An effective learning experience is one which results in a maximum number of desirable changes in the behaviour of the learners.

7. Extension teaching must accomplish certain kinds of educational changes in relation to the subject matter taught: Desirable changes in the knowledge, skill, attitude, understanding, goal, action and confidence of the people are to be achieved in relation to the topic being taught.

8. Extension teaching requires careful evaluation of results: Whether the teaching has been able to attain the objectives set forth *i.e.* desirable changes in behaviour of the people have taken place, have to be assessed or evaluated. Steps are to be taken to meet the deficiencies identified in the process.

Criteria for Effective Leaning

Learning involves acquisition of knowledge, skill, attitude etc., retention to stop reversion and transfer, to use it in real life situations. Learning to be effective and it should have the following characteristics:

1. Learning should be purposeful.
2. Learning should involved appropriate activity by the learners that engages a maximum number of senses.
3. Learning must be challenging and satisfying.
4. Learning must result in functional understanding.

Principles of Learning

The principles of learning may provide good guidance for making learning in extension effective. There are some principles of learning which are very well applicable in extension.

1. Principle of Self-Activity: Learning engaged the maximum number of senses. Learning is an active process on the part of the learners. The kind of learning which takes place is the result of the kind of experience are acquires. The experiences must be compelling and direct on the part of the learners

Conducting demonstration by the farmers in their own fields provides opportunity of self activity *i.e.* learning by doing. This makes learning effective and permanent.

2. Principle of Association: Learning is continuous and therefore, you must know the state of the learners' minds and capacity, and begin teaching at their level for better learning. New learning may be associated with previous successful and satisfying responses. If the farmers have obtained profitable return by the application of nitrogenous fertilizer, they may be motivated to use balanced fertilizer containing phosphate and potash, for still higher return.

3. Principle of Transfer: Application of perceived relationship to another situation in which it is applicable. Unless knowledge or learning can be applied in a new situation, it remains very much restricted. If the farmers have learnt the technique of water management in a particular crop, they should be able to use this knowledge in other crops as well. This shall spread the effect of learning.

4. Principle of Disassociation: For effective learning, undesirable responses are to be eliminated. This may be done by setting up desirable substitutes which are more satisfying. When planting a crop in lines gives better yield, the farmers may be advised not to practice broadcasting.

5. Principle of Readiness: Learning takes place more effectively when one is ready to learn. When farmers are ready to cooperate, with good guidance, they may be able to form a cooperative society.

6. Principle of Set or Attitude: An unfavourable attitude or set retards learning and a favourable attitude accelerates it. Unless attitude become favourable, adoption will not take place.

When farmers develop a favourable attitude towards scientific treatment of diseases in cattle, they shall learn the importance of this type of treatment for animals.

7. Principle of Practice: Learning must result in functional understanding. Perfection is seldom achieved without practice. The practice must be correct, otherwise there will be wrong learning. The attainment of perfection demands that undesirable and useless movement are replaced by desirable and useful ones.

Learning to use a sprayer correctly, requires practice several times over.

8. Principle of Motivation: Motivation or drive means stimulation towards action. Without motivation an organism does not behave and hence does not learn. The practice recommended must be motivating for learning to take place.

Favourable experience of planting trees motivate tribal farmers to collect saplings from the forest nursery.

9. Principle of Timing: Learning is a gradual process and needs multiple exposure for change to occur. Other things being equal, learning takes place more readily when there is introduction of a topic or skill at a time when it can be used in some serviceable manner.

When insects have appeared or are likely to appear in crops, farmers shall readily learnt about plant protection.

10. Principle of Clarity of Objectives: The objective of learning should be clear and purposeful. The ease of learning seems to vary directly with the meaningfulness of the material presented. Meaningful learning is interesting and easier than senseless learning.

When farmers use crop loan only for growing crops, they are clear about the objective of getting the loan. This clear understanding enable the farmers to learn about proper utilization and repayment of loans and take further loans if necessary, for economic development.

11. Principle of Satisfyingness: Learning must be challenging and satisfying. Crops grown during the rabi-summer season give higher economic return and higher level of satisfaction to the farmers. Farmers learnt to invest more and take more care for crops during the season.

In addition to understanding the principles underlying how adults learn, training must also plan experience and activities that maximize the learning of trainees. Edger Dale's 'Cone of Experience', as modified by Sheal (1989), provides a linkage between learning, activity and participant involvement as shown below:

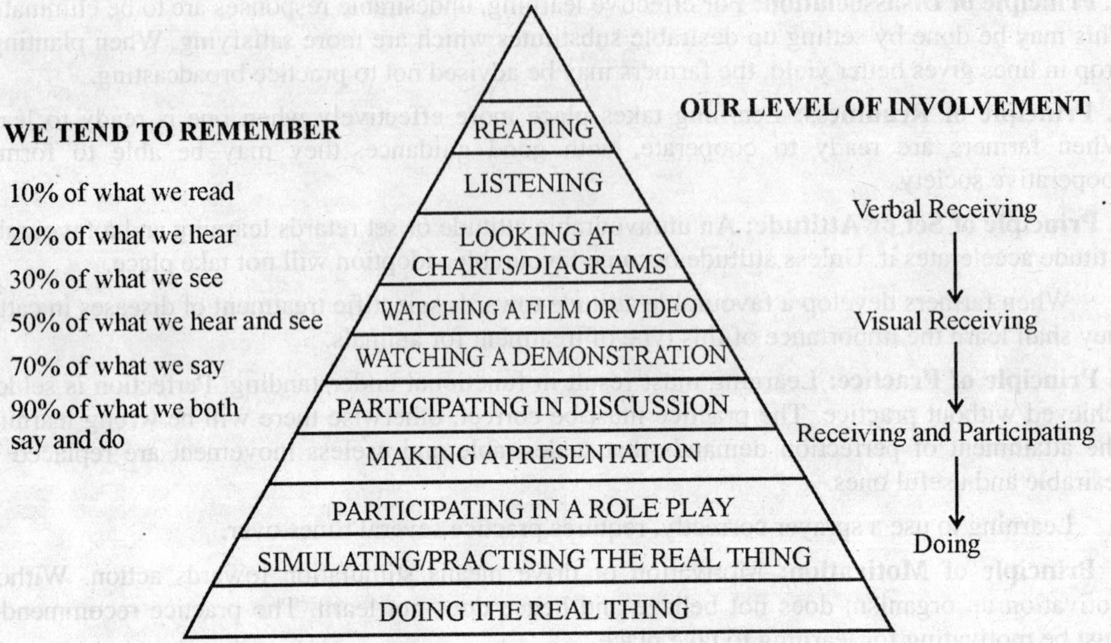

Steps in Extension Teaching

Extension teaching is a planned and deliberate act on the part of the extension agent. The extension agent has to move step to step in a scientific and logical way to impart training to the clients who are farmers, farm woman and rural youth. The role of the extension agents is that of a facilitator and motivator. There are following steps in extension teaching.

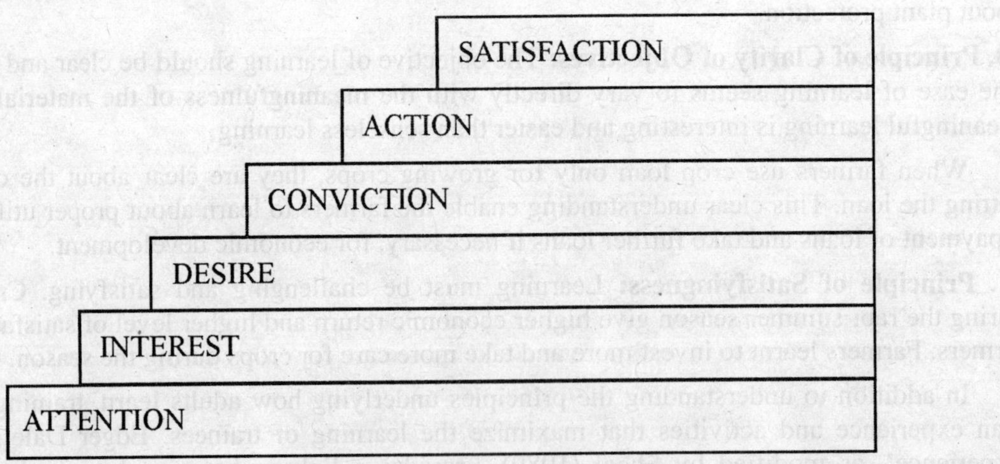

Teaching-learning is a continuous process consisting of various steps. It is difficult to separate steps from one another. According to Wilson and Gallup the following are the steps in teaching-learning process are AIDCAS.

Attention – Interest - Desire - Conviction - Action - Satisfaction

1. **Attention:** The first task of the extension worker is to attract attention of the learners to the new and better ideas. Farmers are to be made aware of the improvement.
2. **Interest:** Once attention has been captured it becomes possible for the teacher to appeal to the basic needs or urges of the individual and arouse his interest in further consideration of the idea. Extension worker reveals how new practice will contribute to the farmer's welfare. The message should be presented attractively.
3. **Desire:** The desire is concerned with continuing farmer's interest in the idea or better practice until interest becomes a desire or motivating force. The extension worker explains the farmer that the information applies directly to the farmer's situation and doing of this would satisfy his needs.
4. **Conviction:** Action follows desire, conviction of the people and prospects of satisfaction. In this step, the learner knows what action is necessary and just how to take that action. He also makes sure that the learner visualizes the action in terms of his own peculiar situation and has acquired confidence in his own ability to do things.
5. **Action:** Unless conviction is converted into action the efforts are fruitless. It is the job of the extension worker to make it easy for the farmers to act. If new control measure is action oriented, the recommended chemical should be available within the farmers reach. Necessary equipments should also be available. If action does not quickly follow the desire the new idea will fade away. Therefore this phase should never be neglected.
6. **Satisfaction:** This is the end product of the process. Follow-up by the extension worker helps the farmers to learn to evaluate their progress and strengths. Satisfaction helps to continue his action with increased satisfaction. Satisfaction is the motivating force for further learning. "A satisfied customer is the best advertisement" will also apply to the extension worker.

The above six steps are often blended with each other and lose their clear cut identity. Of course these steps are based on motivation.

Competencies Needed by the Extension Agent

Competency refers to the ability of an individual to reach a desire goal. These are:

1. Technical competency.
2. Economic competency.
3. Scientific competency.
4. Occupational competency.
5. Communication competency.
6. Social competency.

EXTENSION TEACHING METHODS

Extension teaching methods are the tools and techniques used to create situations in which communication can take place between the rural people and the extension professionals. These are the methods of imparting new knowledge and skills to the rural people by drawing their attention towards such technologies, thereby arousing their interest and helping them to have a successful experience of the new practice. A proper understanding of these methods and their selection for a particular type of work is necessary.

Classification of Extension Teaching Methods

(A) According to Use

One way of classifying the extension methods is according to their use and nature of contact *vis-a-vis* interpersonal, group and mass communication. Based upon the nature of contact, they are classified into following categories:

1. Individual-Contact Methods

Extension methods under this category provide opportunities for face-to-face or person-to-person contact between the rural people and the extension professionals. These methods are very effective in teaching new skills and creating goodwill between farmers and the extension professionals.

2. Group-Contact Methods

Under this category, the rural people or farmers are contacted in a group which usually consists of 20 to 25 persons. These groups are usually formed around a common interest. These methods also involve a face-to-face contact with the people and provide an opportunity for the exchange of ideas, for discussions on problems and technical recommendations. In this way, the future course of action is finalized.

3. Mass or Community-Contact Methods

An extension professional has to approach a large number of people for disseminating information and helping them to use it. This can be done through mass-contact methods conveniently. These methods are more useful for making people aware of the new technologies quickly.

Classification of Extension-Teaching Methods according to their use

Individual contacts	Group contacts	Mass contacts
Farm and home visits	Method and result demonstration	Print media (*viz.* Newspapers, magazines, leaflets, posters, pamphlets, circular letters bulletins)
Office calls	Group meetings, discussions	Electronic media (*viz.* Radio, television, cell-phones)
Telephone calls	Conferences, seminars, workshops, etc.	Internet-based media
Personal letters	Field trips, field days, campaign	Exhibitions, Kisan mela, Gosthi

(B) According to Form

Extension teaching methods are also classified according to their forms, such as written, spoken and audio-visual. Some of the important methods under each of these 3 categories are listed below:

Classification of Extension-Teaching Methods according to their Form

Written	Spoken	Object or visual
Bulletins	General and special meetings	Result demonstrations
Leaflets, folders, news articles	Farm and home visits	Demonstration plots
Personal letters	Official calls	Motion-pictures or movies, charts
Circular letters	Telephone calls, radio	Slides and film-strips, models, exhibits

A brief description of some of the extension methods which are commonly used by extension professionals is given below:

1. Farm and home visit

Farm and home visit constitutes the direct or face-to-face contact by an extension professional with the farmer or the members of his family. During these visits, information is exchanged or discussed. The visits may be to get acquainted with the problems of the farmers. Such visits provide an opportunity for a two-way communication.

***2.* Method demonstration**

It is used to show the technique of doing things or carrying out new practices e.g. milk production, paneer making, ghee making, etc. This method is usually used for groups of people.

***3.* Result demonstration**

Result demonstration is meant for proving the advantages of recommended practices and to demonstrate their applicability to the local conditions. It is conducted by a farmer under the direct supervision of an extension professional. It is designed to teach others, in addition to the person who conducts the demonstration. It helps the farmers to learn by seeing and doing. This method can be used to show the superiority of practices, such as traditional culture, value addition, hygienic handling of animals, etc.

***4.* Group discussion**

All the farmers cannot be contacted by extension professionals individually because of their large number. It is convenient and feasible to contact them in groups. This method is commonly known as group discussion. It is used to encourage and stimulate the people to learn more about the problems that concern the community through discussion. It is a good method of involving the local people in developing local leadership and in deciding on a plan of action in a democratic way.

5. Exhibition

An exhibition is a systematic display of information, actual specimens, models, posters, photographs, and charts, etc. in a logical sequence. It is organised for arousing the interest of the clientele in the things displayed. It is one of the best media for reaching a large number of people, especially illiterate and semi-literate people. Exhibitions are used for a wide range of topics, such as planning a model village, showing high-yielding breeds of cattle and buffaloes, new agricultural implements and the best products of village industries.

6. Campaign

Campaign is used to focus the attention of the people on a particular problem, *e.g.* milk adulteration, vaccination and prevention of animal diseases, pest control etc. Through this method, maximum number of farmers can be reached in the shortest possible time. It builds up community confidence and involves the people emotionally in a programme.

7. Field tour

Conducted tours for farmers are used to convince them and to provide them with an opportunity of seeing the results of new practices and products, skills, etc. and to give them an idea regarding the suitability and application of these things in their own area. Such tours may also be arranged to enable the rural people to visit places and institutions, such as research institutions, training institutions, agricultural universities, model villages, areas of advanced developments, leading private farms, exhibitions, and agricultural and cattle fairs/dairy mela, kisan mela/krashak gosthi

8. Print media

Newspapers, magazines, bulletins, leaflets, folders, pamphlets and wall news-sheets are another set of mass media for communicating information to a large number of literate people. They are used for communicating general and specific information on a programme of technology or a practice. Small folders, leaflets and pamphlets are used to give specific recommendations about a practice, such as clean milk production, vaccination schedule, detection of milk adulterants, improved package and practices of crops, INM, IPMM, newly released varieties etc.

9. Radio

It is one of the most powerful media of communication. It is a mass medium of communication and can reach a large number of people at any given time involving the least expense. Extension professionals use the radio for communicating information on new methods and techniques, giving timely information about the control of animal diseases such as foot and mouth disease (FMD), animal pests, sowing time, weather forecasts, market news, mandi rates of different crops etc. For this purpose, talks, group discussions, folk-songs, dialogues and dramas are usually broadcast. There are radio programmes broadcast by All India Radio (AIR), FM (frequency modulated) radio, community radio, etc.

10. Television

It combines both audio and visual impact and is very suitable for the dissemination of agriculture and dairy information. It is more useful in teaching to do a specific job. A beginning has been made in India for using this medium for development programmes and it is expected that its use will become more extensive in the coming years. At present, along with the Government-owned channel (Doordarshan), several other private channels are telecasting various kinds of entertainment and developmental programmes to reach the viewers.

11. Leaflet

A leaflet is a single sheet of paper used to present information on only one developmental idea in a concise manner, using simple language.

12. Folder

A folder is a single piece of paper folded once or twice, and, when opened, the material is presented in sequence.

13. Pamphlet

A pamphlet is an unbound single sheet of paper that is printed on both sides, printed in colours with action photographs, giving full information about a topic in greater length than in folders or leaflets.

14. Bulletin

A bulletin is a publication of around 20 pages, with the primary objective of giving complete information which the intended readers can apply to their own situation.

15. Booklet

When the extension material exceeds 20 pages and is less than 50 pages, it is called booklet.

Selection of Extension Teaching Methods

Agricultural extension has its roots in researches related to crop and crop production technologies, but the method of sharing this information with farmers depends on a social sciences process. The major role of extension professionals in this process is to educate the agricultural farmers. Hence, it is his responsibility to provide better learning situations so that learners can learn effectively. No single technique alone is suited to all situations. Therefore, a combination of suitable teaching methods may be employed to do the needful, some of which are being discussed below:

1. Methods for Getting Attention

To provide learning experiences for people, extension professionals must get their clients attention. This is the fundamental stage for successfully making the ideas enter the minds of people, and it requires planned and frequent use of a variety of teaching methods, *viz.* pictures, demonstrations, news, success stories, survey results, slogans, posters, radio talks, cartoons, displays, exhibition, etc.

2. Methods for Developing Interest

The amount and direction of developmental accomplishments of extension teaching are largely determined by the interests of the people. Learning without interest does not take place to any appreciable degree. People who are interested in solving developmental problems acquire more information about them. Interest usually represents the objectives or goals of an individual. Development means comparatively little to people until they are able to connect it to their interests.

The factor of interests tends to control the influence of extension education, because only those people who are interested will take part in developmental activities. The methods that are useful in developing people's interest are: meetings, radio talks, publications, tours, result demonstrations, photographs, charts, etc.

3. Methods for Developing Confidence

Development and maintenance of confidence should be parallel to all other changes in people's attitudes and behaviour. Confidence is the key to achieve results from extension education, because action taken by people is voluntary. Confidence of the client corresponds to the goodwill of the extension professionals. Once broken, confidence is difficult to restore, either in oneself, or in the community as a whole. It is, therefore, necessary for extension professionals to safeguard the factor of confidence in all the steps of teaching. In an ideal extension teaching, confidence should grow in intensity, as desire is created and followed by action and satisfaction. Methods that are useful in developing and maintaining confidence are: economical and practical recommendations, readily adoptable developmental practices, widely seen demonstrations, and personal visits.

4. Methods for Creating Desire

Desire has a definite focus, unlike attitudes and interests which are general, which is specific to the objects to which it is attached. Desire comes only when a developmental idea, or intervention, suggested by extension professionals is considered favourable by people. It may be noted here that desire is an outgrowth of interest and confidence.

In creating desire, appeal to the feelings and emotions of people should be employed to support the interest they already have. A desire is a want, and to make that want to be felt and to satisfy the want, the right teaching methods must be used to stimulate feelings and emotions, *viz.* exhibitions, demonstrations, pictures, working models. For example: solar water heating devices, circulars suggesting benefits, 'before' and 'after' pictures that demonstrate development, and actual development results presented in all possible ways.

5. Methods to Ensure Action

Action follows desire. If action does not follow soon after the desire has been created, the desire soon fades away and people continue as before. Consequently, a variety of extension teaching methods must be planned and used to ensure action that result in development. The methods useful for promoting action are: use reminders; ensure the participation of leaders in development work; cooperative action; success stories and radio talks about and by people taking action.

6. Methods for Maintaining Satisfaction

Satisfaction depends upon the confidence, pride and success as a result of adoption of developmental initiatives. After a developmental practice has been adopted by people, it is important to remain in touch with them till they experience real satisfaction and continue with the practice. Continuation of the practice and recommendations to others is an indication of satisfaction. Methods that are useful in maintaining satisfaction are: personal contact, whenever possible; timely hints; showing the value of results; giving more information and publicity in media.

Factors for Selection of Extension Teaching Methods

No single "rule-of-thumb" can be given for the selection and use of the various extension methods for ensuring success in all situations. On one hand, the individual contact methods furnish the most direct opportunities for influencing people effectively; on the other hand all other methods of group and mass procedures are dilutions or compromises created by the pressure of necessity. We must reach more people, teach them more often and keep down the cost per contact.

In order to get most effective results the extension professionals should: select the appropriate methods; have a suitable combination of the selected methods and; use them in proper sequence, so as to have repetition in a variety of ways. For doing this, a number of factors should be considered, such as:

1. Audience

Based upon the principle of *sine qua non*, it is imperative that the instructors must be aware of the type of audience. In addition, size of the audience coupled with level of their educational qualification also contributes significantly as a factor influencing the choice of extension teaching methods. For instance, group discussion cannot be used effectively when the number of participants exceeds thirty; method demonstration can be used for a relatively small audience, while lecture meetings can be used for large audiences.

2. Teaching Objective

Before starting the selection process for identifying suitable methods, the objectives of the whole process must be very clear in minds of the instructor as well as learners. In this regard, some specific queries need to be addressed: do you want to think about a change in thinking or knowledge or in attitude or feeling or in action or skill? If you want merely to inform or influence a large number of people slightly, you should use mass media. If you want to influence a relatively small number of people to make maximum improvements, resort to individual contact methods. If you want to change attitude or arrive at a consensus of opinion, arrange group discussion or work through village leaders. If you want to teach a skill, use the method demonstration.

3. Subject Matter

Where the new practice is simple or familiar (*i.e.*, similar to those already being followed) the news article, radio or circular letter will be effective, whereas complex or unfamiliar practices will require face-to-face contacts, written materials and audio-visual aids.

4. Stage of Development of Extension Organization

In the initial stages of extension, result demonstrations will be necessary to gain confidence of farmers. But if extension work is already well established and the farmers have confidence in extension services, result demonstrations may not be necessary and local illustrations of adoption by village leaders will suffice.

5. Availability of Communication Media

The availability of certain communication media (*i.e.* news papers, telephones, radio etc.) will also have a direct bearing on the extent to which these methods can he used.

6. Relative Cost of the Method

The relative cost of the method (*i.e.*, the amount spent on extension teaching in relation to the extent of practices changed) is also an important consideration in their selection and use.

7. Extension Professionals Familiarity with Extension Methods

An extension professionals familiarity with, and skill in the use of, several extension methods will also influence his choice and use of the methods.

Considerations in the Selection of Extension Teaching Methods

The following guidelines are helpful in the selection of suitable extension teaching methods:

1. Educational Level of the Audience

- For illiterates - Personal visits.
- For educated - Written materials.

2. Size of the Audience

- For less than 30 - Lecture, Group discussion.
- For more than 30 - Mass methods.

3. Teaching Objective

- To bring awareness - Mass methods.
- To change attitude - Group discussion.

- To impart skill - Demonstration.

4. Subject Matter

- To prove value of a recommended practice - Result demonstration.
- To teach a new skill, or an old one in an improved way
- To disseminate simple technology - News article.
- To teach a complex technology - Face-to-face contact with audio visual aids.

5. Extension Organization's Credibility

- New organization, yet to gain confidence of people
- Well established organization with proven success - Circular letter.

6. Size of Extension Staff

- Few staff members - Group and mass contact methods.
- Large number of staff - Individual contact methods.

7. Availability of Media

- For creating awareness and reinforcement of ideas - Television, radio, News paper.

8. Time of Dissemination

- Emergency for individual - Phone call.
- Emergency for a group of people, or a large number of people - Radio, television, public address system.

Chapter - 6

Communication

"Communication is the process of effecting an interchange of understanding between two or more people".

- Teaching and learning is the process of communication.
- Communication is a two way or double way process.
- Teachers acts as catalyst.
- Communication involves:
  a. Source and his message
  b. Receiver and his response.
- Communication means to establish commonness between sender and receiver of message is termed as communication.
- Communication means sharing of ideas and feelings in a mood of mutuality.
- The chief emphasis in communication of target audience (not on material or medium).

Meaning of Communication

- The word 'communication' comes from the Latin 'communis' meaning 'common'. This implies that when we communicate. We are trying to establish 'communality' with someone through a message.
- Communication is a conscious attempt to establish commonality over some idea, fact, feelings and the like with other. In essence, it is a process of getting a source and a receiver tuned together for a particular message or series of message.
- The word communication has many different meanings for different persons. According to Webster's new world dictionary communication means, transmitting, giving and receiving of information, signals or messages by talk, gestures, writing etc.
- The concept of communication as a science or art of communicating as a branch of study and that too in relation to communication in between human beings only.
- Communication making commonness between sender and receiver.
- Interchanging ideas, feelings, facts between two or more people.
- It helps to movement of know-how.
- It is also called learning process.

Definition of Communication

- Communication as the process by which two or more people exchange their ideas, facts and feelings in a way that each gains a common understanding of the meaning, intent and use of message (Legans).
- Communication is a movement of knowledge between two or more personality in order to achieve the expected results (Winfield).
- Communication is the process of interchanging the ideas opinions, experiences between two or more persons (Stevans).
- Communication is a deliberate attempt on the part of the communications in order to felicitate to expected response from the others (Bellinghass).
- Communication is anything that conveys *meaning*, that carries a *message* from one person to another (Broller, 1949).
- Communication is the discriminatory response of an organism to stimulus (Stevans, 1942).
- Communication is the mutual interchange of ideas by any effective means (Thayer, 1968).
- Communication is the control of behaviour through discipline and reinforcing stimuli (Hastman, 1966).
- Communication is the process of effecting an idea interchange of understanding between two or more peoples (Thayer, 1968).
- Communication is the arrangement of environmental stimulus to produce certain desired behaviour on the part of the organisms (Thayer, 1968).
- Communication occurs when two corresponding system coupled together through one or more non-corresponding systems assume identical status as a result of single transfer along the chain, we are trying to established commonness (Schramm).

Communication systems

K - Knowledge
G - Generating
S - System
D - Disseminating
C - Consuming

Agencies

Universities
Research institutes KGS
Private organization
Voluntary organization

Mass media

Universities
Research Institutes KDS
Govt. Agencies
Voluntary Organizations

Farmers

In Agriculture KCS

Key elements of communication

a. Sender
b. Message
c. Message treatment *i.e.*, encoding
d. Communication medium
e. Receiver: It is the most important element because every element is distinct towards receiver
f. Receiver response

Elements of Communication

I. COMMUNICATOR

- In the context of agricultural and rural development extension worker/extension functionaries/change agent is the communicator who starts the process of communication. The extension worker and mass media like radio are sometimes visualised as sources or originators of message, which is not correct. Knowledge generates research through research projects.
- Agricultural Universities are the originators or sources of message. The extension worker obtains the required information from research and carries it to the audience and the farmers. To enhance the process, he may take the help of some aids, known as audio-visual aids. He also carries back the reactions of the farmers, their problems etc. as feedback information to research for finding out solutions for the same.
- The credibility of the communicator and the organisation he represents for effective communication. Credibility means trust worthiness and competence. Before the audience accepts any message will judge whether communicator and the organisation can be relied on and is competent enough to give the information, who tells is, therefore, very important in extension communication.

The characteristics of a good communicator are

The individuals knows

i. His objectives –have them specifically defined.
ii. His audience – their needs, interests, abilities, predispositions
iii. His message – its content, validity, usefulness, importance
iv. Channels that will reach the audience and their usefulness
v. How to organise and treat his message
vi. His professional abilities and limitations.

The individuals is interested in

i. His audience and its welfare
ii. His message and how it can help people
iii. The results of communication and their evaluation
iv. The communication process
v. The communication channels – their proper use and limitations.
vi. How to improve his communication skills.

The individuals prepares

i. A plan for communication – a teaching plan
ii. Communication materials and equipments
iii. A plan for evaluation of results

The individuals has a skill in

- Selecting messages
- Treating messages
- Expressing messages – verbal and written
- The selection and use of channels
- Understanding his audience
- Collecting evidence of results.

Poor communicators on the other hand

a. Fail to have ideas to present that are really useful to the audience.
b. Fail to give the complete story and show its relationship to people's problems.
c. Forget that time and energy are needed to absorb the material presented.

d. Feel they are always clearly understood.
e. Refuse to adjust to close minds.
f. Talk while others are not listening.
g. Audience understanding.
h. Fail to recognise others view point and develop presentation accordingly.
i. Fail to recognise that communication is a two way process.
j. Let their own biases over influence the presentation.
k. Fail to see that everyone understand questions brought up for discussions.
l. Fail to provide a permissive atmosphere.
m. Disregard the values, customs, prejudices and habits of people with whom they attempt to communicate.
n. Fail to start where people are, with respect to knowledge, skill, interest and need.

II. MESSAGE

- A good message should clearly state what to do, how to do, when to do and what would be the result.
- To produce desirable changes in human behaviour, the message must be motivating.
- Messages which are relevant, interesting behaviour, the useful, profitable, credible (Latest and best on research findings) and complete (Neither too much, nor too little) are likely to motivate the people.

A good message must be

a. In line with the objective to be obtained.
b. Clearly understandable by the audience.
c. In line with the mental, social, economic and physical capabilities of the audience.
d. Significant –economically, socially or aesthetically to the needs interests and values of the audience.
e. Specific – no irrelevant materials.
f. Simply stated –covering only one point at a time.
g. Accurate – scientifically sound, factual and current.
h. Timely – Especially when seasonal factors are important and issues are current.
i. Supported by factual material covering both sides of the agreement.
j. Appropriate to the channel selected.
k. Appealing and attractive to the audience-having utility immediate use.
l. Applicable – audience can apply recommendation.
m. Appealing and attractive, adequate – combining principle and practice in effective proportion.
n. Manageable - can be handled by the communicator and within limits of time.

In contrast, poor communication often

i. Fail to clearly separate the key message from the supporting content or subject matter.
ii. Fail to prepare and organise their message properly.
iii. Use inaccurate or fuzzy symbols –words, visual or real objects to represent the message.
iv. Fail to select messages that are sharply in line with the felt needs of the audience.
v. Fail to present the message objectively – present to material, often biased, to support only one side of the proposition.
vi. Fail to view the message from the standpoint of the audience.
vii. Fail to time the message properly within a presentation or within a total programme

III. CHANNEL

- Channel of communication constitutes the medium through which information flows from a sender to one or more receivers. Face to face, word of mouth is the simplest and yet one of the most widely use and effective means of communication, particularly for the developing countries.
- As society changes from traditional to modern, the communication channel changes from oral to media system. Because of the large member of audience or receivers of information and because of physical distance of the communicator and the receivers of information, it is necessary to use different media of communication. Even in interpersonal, face to face, word to mouth communication, it becomes necessary to use some aids to make communication more effective.

The channels of communication may be classified into a number of ways according to different criteria.

According to form

Spoken: Farm and home visit, office call, meeting of all kinds, radio talk etc.
Written: Personal letter, farm publications, news paper etc.

According to nature of personnel involved

Personal localite

- They are the local leaders and local people who belong to receivers own social system. Personal localite channels are important in traditional social system.

Personal cosmopolite

- These are the channels of communication from outside the social system of the receiver. They are the extension workers of various organisations and are important in changing the farmers from traditional to modern.

Impersonal cosmopolite

- When the channels of communication are from outside the social system of the receiver and at the same time no personal face to face contact is involved. These are mass media, which are important in areas of high urban influence and farmers who are modern or are changing from traditional to modern.

According to nature of contact with the people

Individual contact

- The extension worker communicates with the people individually, maintaining separate identity of each person.
- Examples are farm and home visit, office call, personal letter etc.

Group contact

- The extension worker communicates with the people in groups are not as individual persons.
- Examples are group meeting, small group training, field day or field visit, study tour etc.

Mass contact

- The extension worker communicates with a mass of people without taking into consideration, their individual or group identity.
- Examples are mass meeting, campaign, exhibition, radio, television, smart phone, Webnar, etc

IV. TREATMENT OF MESSAGE

- Treatment means the way a message is handled, so that the information gets across the audience. It relates to the techniques or details of procedure or manner of performance, essential to effective presentation of the message clear, understandable and realistic to the audience.
- Treatment of the message by the communicator shall depend to a great extent on choice of the channel and the nature of audience. The task cannot be reduced to a formula or recipe. Treatment is a creative task that has to be tailor-made for each communication function.
- For example, treatment of a message will be different to when it is conveyed in a meeting or published in a folder or telecast. Similarly, there will be difference in the treatment of the message according to the level of literacy, socio-economic condition and degree of progressiveness of the audience.

- Designing treatments usually requires original thinking, deep insight into the principles of human behaviour and skill in creating and using refined techniques of message presentation.

The following are the categories on the bases of useful for varying treatment.

Matters of general organisation

1. Repetition or frequency of ideas and concepts.
2. Contrast of ideas.
3. Chronological compared to logical and psychological.
4. Presenting one side compared to two sides of an issue.
5. Emotional compared to logical appeals.
6. Starting with strong arguments compared to them until the end of presentation.
7. Inductive compared to deductive.
8. Proceeding from the general to the specific and vice versa.
9. Explicit drawing conclusions compared to leaving conclusions implicit for the audience to draw.

Matters of speaking and acting

1. Limit the scope of presentation to a few basic ideas and to the time allotted. Too many ideas at one time are confusing.
2. Be yourself, you can't be anyone else. Strive to be clear, not clever.
3. Know the facts, fuzziness means sure death to a message.
4. Don't read your speech. People have more perfect for a communicator who talks to the audience.
5. Know the audience. Each audience has its own personality, be responsive to it.
6. Avoid being condescending (Patronising). Do not talk or act down to people, or over their heads. Good treatment of messages result in hitting the target. Never overestimate the knowledge of an audience or underestimate its intelligence.
7. Decide on the dramatic effect desired. Effective treatment required sincerity, smoothness, enthusiastic, warmth, flexibility and appropriateness of voice, gestures, movements and tempo.

V. AUDIENCE

- The audience or receiver of message is the target of communication function. An audience may consist of a single person or a number of persons. It may comprise men, women and youth. An audience may be formed according to occupation groups such as field crop farmers, fruit growing farmers, dairy men, poultry keepers, fish farmers, homemakers etc. Audience may also be categorised according to farm size such as marginal, small, or large farmers or according to farm size whether belong to scheduled caste, scheduled tribe etc.

- In addition to knowing the identity of an audience and some of its general characteristics, there are other somewhat more specified aspects that help to classify the exact nature of an audience and how to reach it.

The following are some of these.

1. Communication channels established by the social organisation.
2. The system of values held by the audience-custom, tradition etc.
3. Forces influencing group conformity- custom tradition etc.
4. Individual personality factors – susceptibility to change etc.
5. Native and acquired abilities.
6. Educational, economic and social levels.
7. Pressure of occupational responsibility how busy or concerned they are.
8. People's needs are as they see them, and as the professional communicators see them.
9. Why the audience is in need of changed ways of thinking, feeling and doing.
10. How the audience views the situation.

VI. AUDIENCE RESPONSE

The response of the audience is the ultimate objective, the goal of any communication function. Response of an audience to messages received may be in the farm of some kind of action results, extension communication does not achieve its most essential objective.

- Understanding versus knowledge
- Acceptance versus rejection
- Remembering versus forgetting
- Mental versus physical action
- Right versus wrong

VII. FEEDBACK

Extension communication is never complete without feedback information. Feedback means carrying some significant responses of the audience back to the communicator. Communication work is not an end itself. The extension worker *i.e.* the communicator, should know what happened to the audience, the farmers, after the message has reached.

Characteristics of the feedback

- Feedback is source oriented.
- Feedback varies in different communication situations.
- Feedback affects the source or communicators.
- Feedback exerts control over future messages.

- Feedback affects communication fidelity.
- Feedback maintains the stability and equilibrium of a communication system.

Basic functions of communication

Communication has four basic functions. These are:

1. **Information functions:** The basic requirement of adopting and adjust/oneself to the environment is information.
2. **Command or instructive function:** Those who are hierarchically superior. In the family, society or organization, often initiate communication either for the purpose of informing their subordinates or for the purpose of telling them, what to do, how to do, when to do etc. The command and instructive functions of communication are more observable in formal organization than in informal organization.
3. **Influence or persuasive functions:** The sole purpose of communication is to influence people. Persuasive function of communication *i.e.* to induce people, is extremely important for extension in changing their behaviour in the desirable direction.
4. **Integrative function:** A major function of communication is integration or of continuously offsetting any disintegration at the interpersonal or at the organizational level.

Models of communication: Models of communication are conceptual models used to explain the communication process.

Communication code scheme

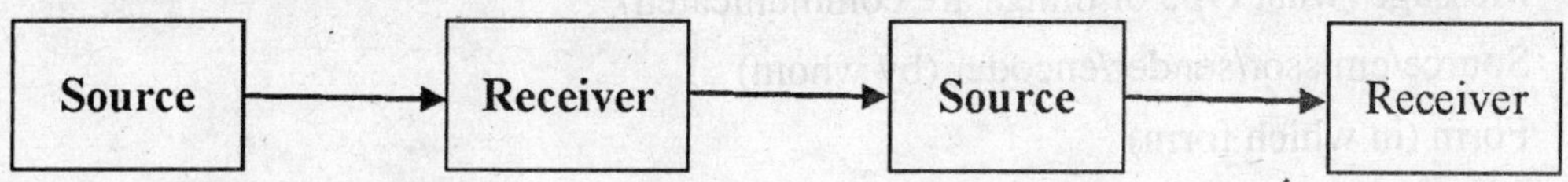

Linear communication model

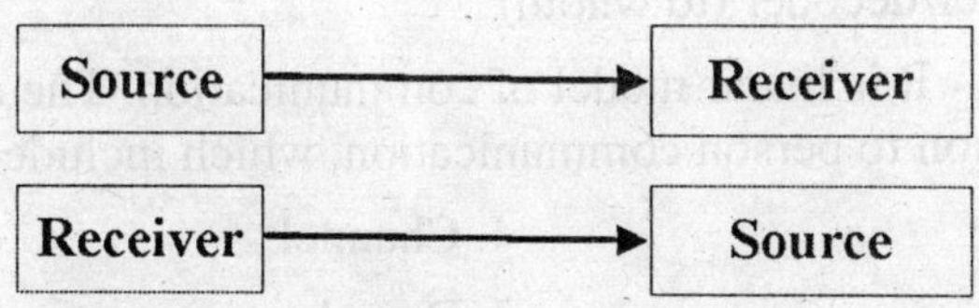

Interactional model of communication

S	M	C	R
Communication skills Attitude Knowledge Social system Culture	Content Elements Treatment Structure Codes	Hearing Seeing Touching Tasting Feeling	Communication skills Attitude Knowledge Social system Culture

According to Aristotle: communication has three ingredients:

1. Speaker – the person who speaks
2. Speech – the speech that the individual produces
3. Audience – the person who listens

According to Shannon – Weaver (1949)

1. Source – which produces a message.
2. Transmitter – which encodes the message into signals.
3. Signal (channel)– which signals are adapted for transmission.
4. Receiver – which reconstructs the encoded message from a sequence of received signals and decodes it.
5. Destination

According to Leagans (1963)

1. Communicator
2. Message
3. Channel
4. Treatment
5. Audience
6. Audience response

According to Schramm (1954)

1. Message (what type of things are communicated)
2. Source/emissor/sender/encoder (by whom)
3. Form (in which form)
4. Channel (through which medium)
5. Destination/receiver/decoder (to whom)

According to Berlo (1960) – It is linear model of communication. The communication process is a simple application for person to person communication, which include:

1. Source
2. Encoder
3. Message
4. Channel
5. Decoder
6. Receiver

According to Rogers and Shoemaker (1971)

1. Source
2. Message
3. Channel
4. Receiver
5. Effects

According to Barnlund (1970)- proposed a transactional model of communication

1. In a slightly more complex form, a sender and a receiver are linked reciprocally.

2. Attitude of communication, referred to as the constitutive model of constructionist view, focuses on how an individual communicates as the determining factors of the way the message will be interpreted.

Classification of communication methods

The choice of a channel or method of communication, also known as extension teaching method, generally depends on the number and location of the target audience and the time available for communication. The extension agent has to choose a particular method or combination of methods according to the needs of the situation.

The communication methods adopted in extension may be classified as follows:

A. INDIVIDUAL METHODS

In this method, the extension agent communicates with the people individually, maintaining separate identity of each person e.g. farm and home visit, farmer's call etc.

Advantages:

Individual method has following advantages. These are:

1. Helps the extension agent in building rapport.
2. Facilitates gaining first have knowledge of farm and home.
3. Helps in selecting demonstrators and local leaders.
4. Helps in changing attitude of the people
5. Helps in teaching complex practices.
6. Facilitates transfer of technology.
7. Enhances effectiveness of group and mass methods.
8. Facilitates getting feedback information.

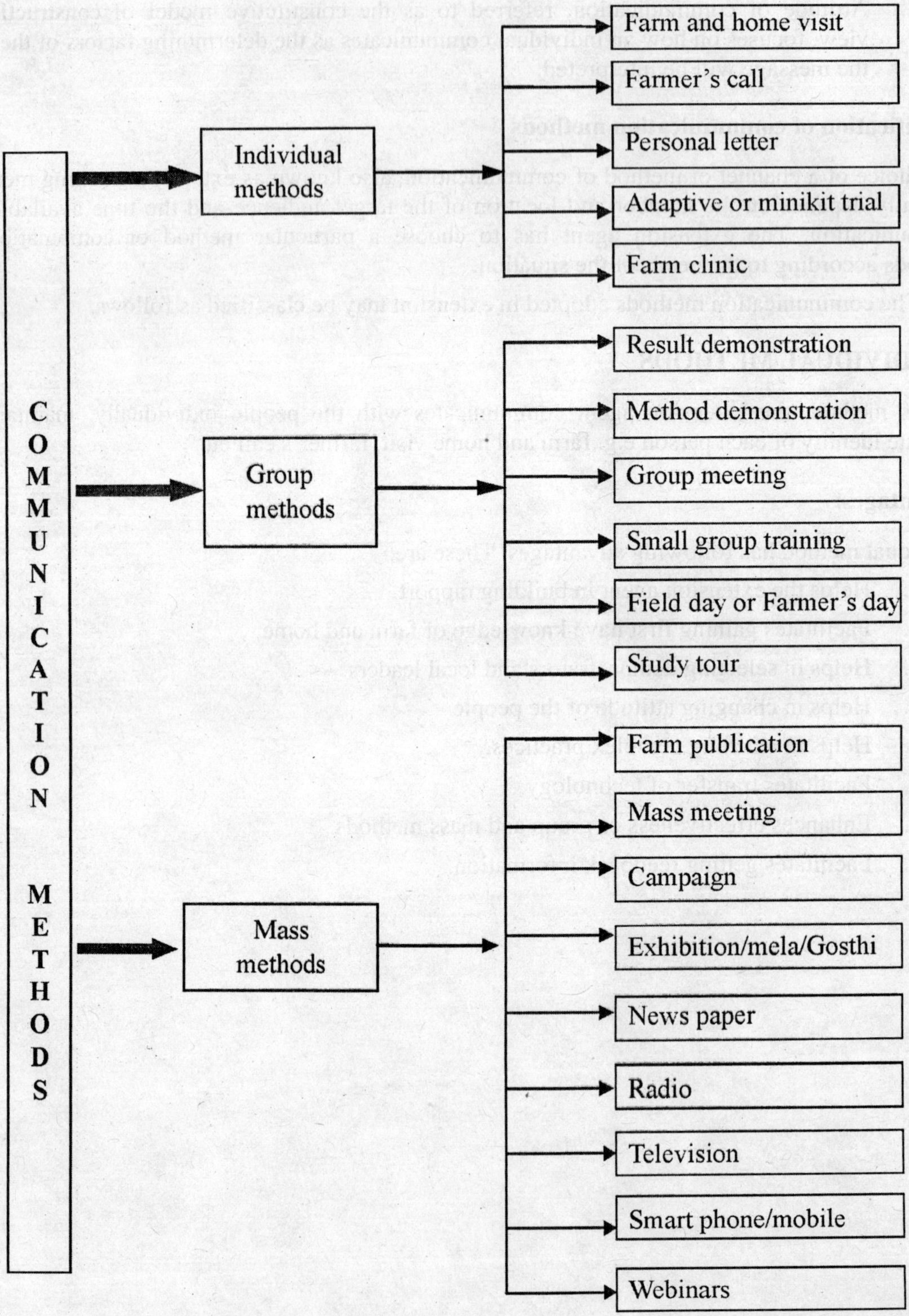
COMMUNICATION METHODS
Individual methods
Farm and home visit
Farmer's call
Personal letter
Adaptive or minikit trial
Farm clinic
Group methods
Result demonstration
Method demonstration
Group meeting
Small group training
Field day or Farmer's day
Study tour
Mass methods
Farm publication
Mass meeting
Campaign
Exhibition/mela/Gosthi
News paper
Radio
Television
Smart phone/mobile
Webinars

Limitations

1. This method is time and labour expensive
2. Low coverage of audience
3. Extension agents may develop favouritism or bias towards some persons.

1. **Farm and home visit:** Farm and home visit is a direct face-to-face contact by the extension agent with the farmer or home maker at their farm or home for extension work.

2. **Farmer's call:** Farmer's call is a call made by a farmer or homemaker at the working place of the extension agent for obtaining information and assistance.

3. **Personal letter:** Personal letter is written by the extension agent to particular farmer or home maker in connection with extension work. This should not be regarded as a substitute for personal contact.

4. **Adaptive or minikit trial:** Adaptive or minikit trial is a method of determining the suitability or otherwise of a new practice in farmers situation.

5. **Farm clinic:** Farm clinic is a facility developed and extended to the farmers for diagnosis and treatment of farm problems and to provide some specialist advice to individual farmers.

B. GROUP METHODS

A group may be defined as an aggregate of small number of people in reciprocal communication and interactions around some common interest. In this method, the extension agent communicates with the people in groups and not as individual persons.

In this situation, these may be a few communicators such as the extension agent and some subject matter specialist. The size of a small group may be from 15 to 25, a medium group from 25 to 30 and a large group from 50 to 100 persons. Group methods are result demonstration, method demonstration, group meeting, small group training, field day or farmer's day, study tour etc.

Advantages

1. Enables the extension agent to have face to face contact with a number of people at a time.
2. Can reach a selected peoples of the target group.
3. Facilitates sharing of knowledge and experience and thereby strengthen learning of the group members.
4. Reach fewer people, but offer more opportunities for interaction and feedback.
5. Satisfies the basic urge of people for social contacts.
6. Motivate people to accept change due to group influence.
7. More effective than mass method in stimulating action.
8. Less expensive than individual method due to more coverage.

Limitations

1. Wide diversity in the interest of group members may create a difficult learning situation.
2. Holding the meeting may be regarded as an objective in itself.
3. Vested interest, caste groups and village fraction may hinder free interaction and decision making by the group members.

1. Result demonstration: Result demonstration is a methods of motivating the people for adoption of a new practice by showing its distinctly superior result.

2. Method demonstration: A method demonstration is given before a group of people to show how to carry out an entirely new practice or an old practice in a better way. It involves seeing, hearing, participating and practicing in a group which shall stimulate interest and action.

3. Small group training: Small group training is a technique of imparting specific skills to a group of people who need them by creating an appropriate learning situation. This is an effective method for transfer of technology.

4. Group meeting: Group meeting is a method of democratically arriving at certain decision by a group of people, by taking into consideration the members points of view. The convenient size of group for conducting a meeting or discussion may be around 15 to 25, which may be extended upto about 50.

5. Field day or farmer's day: Field day or farmer's day is a method of motivating the people to adopt a new practice by showing what has actually been achieved by applying the practice under field conditions. A field day or farmer's day may be held in a research farm or farmer's field or at home.

6. Study tour: In study tour, a group of interested persons accompanied and guided by one or more extension agents moves out of their neighbourhood to study and learn significant improvements in farm and home. The main purpose is to motivate the visitors by showing what others have been able to achieve.

C. MASS METHODS

In this method, the extension agent communicates with a vast and heterogeneous groups of people, without taking into consideration their individual or group identity. This method is followed where a large and widely dispersed audience is to be communicate within a short time. The size of audience may be a few hundreds in mass meeting, few thousands in campaign and exhibition and millions in newspaper, radio and television, smart phones/mobile and webnar.

Advantages

There are several advantages of mass method. These are

1. Suitable for creating general awareness amongst the people.
2. Helps in transferring knowledge and farming and changing opinions.
3. Large number of people may be communicated within a short time.

4. Facilitates quick communication in times of emergency.
5. Reinforces previous learning.
6. Less expensive due to mass coverage.

Limitations

1. Less intensive method.
2. Little scope for personal contact with the audience.
3. Little opportunity for interaction with and amongst the audience.
4. Generalized recommendations hinder application by individuals.
5. Little control over the response of the audience.
6. Difficulty in getting feedback information and evaluation of results.

1. **Farm publication:** Farm publication is a class of publications prepared by the extension agency in printed farm, containing information relating to the improvement of farm and home. Farm publications are of various types such as leaflet, folder, bulletin, news letter, journal and magazine. Farm publications are extremely useful to the literate farmers.

a. **Leaflet:** It is a single printed sheet of paper of small size, containing preliminary information relating to a topic. It is made as when needed. Generally distributed free of cost.

b. **Folder:** It is a single printed sheet of paper of big size, folded once or twice and gives essential information relating to a particular topic. It is printed as and when required. Generally distributed free of cost.

c. **Bulletin:** It is a printed, bound booklet with a number of pages, containing comprehensive information about a topic. It is a printed as and when required. A small price may be fixed.

d. **Newsletter:** It is a miniature newspaper in good quality paper, containing information relating to the activities and achievements of the organization. It has a fixed periodicity of publication. Generally distributed free of cost.

e. **Journal and magazine:** These are periodicals, containing information related to various topics of interest not only for the farmers but also for the extension agents and researchers. It has a fixed periodicity of publication. Generally supplied against pre-payment of subscription for a particular period.

2. **Mass meting:** Mass meeting is held to communicator interesting and useful information to a large audience at a time. The size of the audience for mass meeting may be a few hundreds, but at a time of fairs or festivals it may be few thousands. Mass meeting may be held in a covered or in an open place.

3. **Campaign:** A campaign is an intense educational activity for motivating and mobilizing a community to action, to solve a problem or satisfy a need urgently felt by it. The duration of a campaign may be for a single day on them like 'water for life', for few weeks as in 'rat control' or 'family planning', for a few months as in 'van-mahotsava' and for a few years as in 'grow

more food' campaign. The campaign may be organized once or may be repeated year after year. It may be held by involving a small number of people in a few village.

4. **Exhibition:** An exhibition is a systematic display of models, specimens, charts, photographs, pictures, posters, information etc. in a sequence around a theme to create awareness and interest in the community. This method is suitable for all types of people. Exhibition may be held at the village, block, sub-division, district, state, national and international levels. Exhibitions may also be organized by taking advantages of local fairs and festivals.

5. **Newspaper:** Newspaper is a bunch of loose printed papers properly folded, which contains news, views, advertisements etc. and is offered for sale at regular intervals, particularly daily or weekly. Newspapers are usually printed on a special type paper, known as newsprint. Daily newspapers are resource strong and are published from national/state capitals or big cities. Small news papers are resource weak and are published at weekly or fortnightly intervals, generally from the district or subdivisional headquarters. The circulation of daily newspapers may range from a lakh to several lakhs while small newspapers is limited to a few thousands only.

6. **Radio:** Radio is an electronic audio-medium for broadcasting programmes to the audience. Radio is suitable for creating general awareness amongst the people, help change their attitude and reinforce learning. The medium is extremely convenient for communication in times of crises and urgent situations. This medium is cosmopolite in approach and is suitable for communication to million of people widely dispersed and situated in remote areas. Availability of low cost transister sets has helped radio to penetrate deep knowledge into the rural life.

7. **Television:** Television is an electronic audio-visual medium which provides pictures with synchronized sound. Television combines the immediacy of radio with the mobility of cinema and can carry message over long distances at a relatively low unit cost. Television is a multi-media equipment as it can include motion picture, recording, slide, photograph, drawing, poster etc. Television can show taped as well as live programme.

8. **Mobile phone/mobile:** In today's world, almost everybody owns a mobile phone. This huge reach, if harnessed in agricultural extension, can change the face of agriculture altogether in a developing country like India where we have nothing to lose by using it as a medium to disseminate agricultural information in multimodel form. Many initiatives have been taken in this regard to utilize mobile phones by private sector, public sector, ICAR, SAUs, Govt. of agriculture, IMD, in agricultural advisory service for agronomic practices, weather forecasts and market price. With increasing dependency, the mobile phone is becoming a common communication platform of the world, especially for agriculture.

9. **Webinars:** e-Agriculture organizes regularly webinars to keep our members informed about important developments, emerging trends innovations or strategic issues within the ICT for agriculture sector. To organize webinars, e-agriculture works with partner organizations or with members from the COP.

■■■■

Chapter - 7

Communication Methods

- Most of the times, when the word communication comes to the mind of the people, they think about exchanging ideas and information by means of words but such type of verbal communication is just a small component of communication.
- In the 21st century, businesses have access to a number of methods of communication which can be used with internal and also with external audiences.
- The main methods of communication which the businesses can use include the traditional business meetings as well as print to the video conferencing and social media of the new age.
- Several tools can be utilized by the businesses for addressing audiences who are in locations which are both near and far.
- While a lot of attention is received by the methods of effective communication that are new, the value along with the relevance of the traditional methods of communication still exists.
- Thus, several different methods of communication exists and the determination of the method that is right for the organisation is a vital decision that needs to be taken.
- A connected team and a connected organization in today's world is the only way we can achieve greater things.
- Connections and relations are built with exchange of thoughts and information.
- Communication serves as a valuable mechanism in building these connections and engagement between people.
- Communication is a very important skill and art. There are four different ways we communicate with others. No matter what form of communication we choose, it is important to have the right intention to connect with the other person.
- Communication is a collective term given to the series of steps involved in sending and receiving of information between two or more people.
- The person sending the message is called as sender while the person receiving the message is called as receiver.

Communication may be defined as a two-way process in which participants not only exchange (encode-decode) information, views, ideas and feelings but also create mutual understanding and share common meaning.

Communication is even more valuable in a business environment as there are several parties involved. Various stakeholders, whether they are customers, employees or the media, are always sending important information to each other at all times.

Types of Communication

Numerous new instruments have emerged over the years to help people communicate effectively.

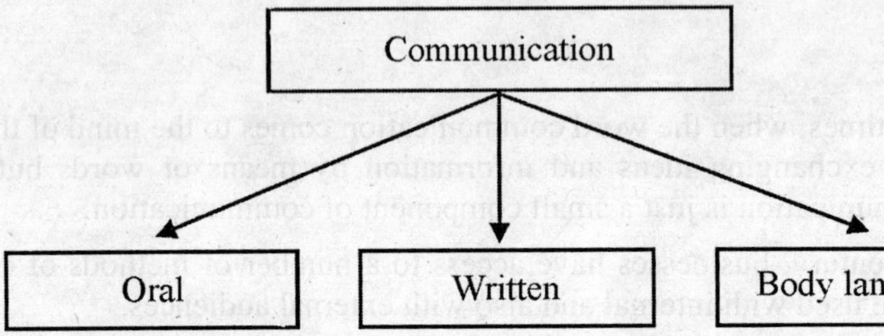

Therefore, it is crucial that different methods of communication are employed.

The methods of communication could be in a sense make or break the management structure of your organization and could also affect the relationship with customers, if not chosen carefully.

The methods of communication can be studied under the following heads:

A. Verbal communication
B. Non-verbal communication
C. Gestural communication (Body language)

Some of the methods of communication are:

A. Oral communication
B. Written communication
C. Kinesics
D. Paralanguage
E. Proximics
F. Signs
G. Actions
H. Silence
I. General behaviour and Attitude
J. Face-to-Face talk
K. Orders
L. Instructions
M. Facial Gestures
N. Communication through body postures

Different occasions require different types of communication. All communication cannot be the same type and it cannot flow through the same means. All depends on who sends the message to whom and for what purposes.

A. Verbal communication

The term verbal communication means 'words'. Human beings are the only species gifted with the use of words that make language. Words are the most accurate and powerful set of symbols. Thus, the common set of symbols are used between sender and the receiver in the process of verbal communication. The word may be spoken or written. The spoken communication is called oral communication. Communication conveyed through written words is called written communication.

Verbal communication is a medium used for transmission of information, views and ideas through use of words. Wherever words are used, spoken or written, for the purpose of conveying something it is known as verbal communication.

Verbal communication may be two types:

1. Oral communication
2. Written communication

(i) Oral communication

- Oral communication could be said to be the most used form of communication. Whether it is to present some important data to your colleagues or lead a boardroom meeting, these skills are vital.
- The person constantly using words verbally to inform subordinates of a decision, provide information, and so on. This is done either by phone or face-to-face.
- The person on the receiving end would also need to exercise much caution to ensure that he/she clearly understands what is being said.
- This shows therefore that you would need to cultivate both your listening and speaking skills, as you would have to carry out both roles in the workplace with different people.

Oral communication is a medium used for transmission of information, views and ideas through use of spoken words. It involves individuals conversing with each other either face-to-face or through some electronic or other channel. Oral communication is normally adopted as a medium to communicate when one intends to communicate instantly and want to build a personal report. Oral messages are temporary in nature as it is difficult to retain them for the future, although with modern electronic devices, now even oral communication can be recorded and saved. The effective of oral communication is influenced by pitch, voice modulation, clarity, volume and speed of spoken words.

It is naturally the speech that precedes the written. In the same way, people speak more before committing themselves into writing. This communication plays an important role in families and organizations and among friends, relatives, acquaintances, neighbours, etc. Miscommunication or ineffective oral communication can trouble the relationship and growth. It may take the form of speeches, discussions, debates, face-to-face talk, telephonic conversation, video conferencing, interview and chat.

It may be formal or informal. Oral communication would be effective if the principle of brevity and conciseness is observed. Mechanical devices used for oral communication include telephone, internet, dictating machines, internal addressing system, internet, etc.

There are various channels of oral communications:

i. Face-to-face conversation
ii. Telephonic conversation
iii. Lectures
iv. Meetings
v. Interviews
vi. Seminars and conferences
vii. Video conferencing

Advantages

i. Speedy communication - when rapid and instant decision are to be taken.

ii. Flexibility- It is flexible as its content can be adjusted and modified according to situation, time frame and listener profile.

iii. Quick feedback and response – It enables the speaker to gather immediate response and feedback from the recipient. Queries, doubt, clarifications, comments, etc., can be given instantaneously.

iv. Economical – Main resource used for oral communication is the spoken words, thus it is economical in terms of saving expenditure on stationary and other resources.

v. Advantageous for large groups - Whenever some common message is to be conveyed to large group of people, then it is best option due to fast, personal and direct.

vi. Personalised touch – The most significant advantage of oral communication is that it attaches personalised touch to the message. With change in voice modulation and volume, sender may bring variations in message according to individuals.

vii. Effective for all – Oral communication is generally speaking a valid mode for all types of listeners, irrespective of their literacy level and capacity to read.

Disadvantages

i. No validity – Oral communication, if not recorded or retained, does not carry any validity or authenticity. One way always deny from the content of oral communication for want of evidence.

ii. Physical distance – If sender and receiver are distantly placed without any facility of mechanical device, oral communication in such situation is not possible.

iii. Scope of misinterpretation – In oral communication, scope of misinterpretation of message is high. This may be due to inattentiveness of receiver, ambiguous speech of speaker or poor verbal ability of speaker.

iv. Unsuitable for lengthy message – Oral communication is not a correct mode of communication when message to be transmitted is very large. In oral transmission, retention is less and there is a possibility of missing on some vital points.

v. Problem in accountability fixation – With oral communication, keeping of documentary proof of the message is not possible. Even a recorded audio may not sufficient as evidence. Thus, if something goes wrong, it is difficult to fix responsibility and hold somebody accountable on the basis of oral communication.

vi. Repetitive viewing not possible – Oral messages cannot be revisited again for the purpose of future analysis or reference, Where the nature of message is repetitive in nature and do not require any additional input, oral communication may prove to be time consuming and burdensome.

(ii) Written communication

Written communication may be defined as a medium for transmitting information and messages with the usage of words written (or typed). Written form of communication may either be in physical form, *i.e.* on paper or may be in electronic form, *i.e.*, e-mails, soft copies on computer, phone messages, etc. Organisations, being formal structures, follow written mode of communication very comprehensively and hence written communication is also termed as business communication.

Writing is used when you have to provide detailed information such as figures and facts, even while giving a presentation.

- It is also generally used to send documents and other important material to stakeholders which could then be stored for later use as it can be referred to easily as it is recorded. Other important documents such as contracts, memos and minutes of meetings are also in written form for this purpose.
- It can be seen in recent years, however, that verbal communication has been replaced to a great extent by a faster form of written communication and that is email.
- It can also be used video conferencing and multiple way phone calls with several individuals simultaneously. Apart from a few glitches that could occur, these methods of communication have helped organizations come a long way.

Message put in black and white constitutes written communication. The written communication is binding on the organization and it constitutes as a documentary evidence. It can be in the form of notices, telex messages, circulars, reports, e-mail and manuals. The components involved in written communication include the writer, the content, the language, purpose of communication, style adopted and the receiver.

Some of the various forms of written communications are:

i. Postal letters

ii. Reports

iii. Journals

iv. Notes

v. E-mails

vi. Magazines

vii. Brochures

viii. Websites
ix. Information booklets
x. Forms

Advantages

i. Permanent record - Written communication can be retained as records for future reference. Such communication being permanent and available for future reference, decisions based on them are more rational as one can review them in detail,

ii. Clarity in presentation – Wherever one intends to transfer numerical data, written communication is the best made. It not presents numerical data in systematic manner, but also complex data having diagrammatic presentation is also captured very effectively.

iii. Reduced possibility of alteration – Once written information is documented and send to the recipient, there is negligible chance that it can be altered and thus it also adds on to the authenticity and reliability of such information.

iv. Useful for repetitive communication: The communication is repetitive in nature, written communication is the most appropriate mode as once designed it can be used again and again with minimal changes. It saves a lot of time, effort and energy.

v. Accountability – Written documents usually carry the name of the sender and are signed. Therefore, such documents attach responsibility of the content on the person sending it.

vi. Effective communication – Message to be sent in written form need not be delivered at the spur of the moment. Instead they can be revised and well edited to the shaped to the maximum effect. Such message are, therefore, more dependable and reliable.

Disadvantages

The limitations and disadvantages of written communication are as follows:

i. Expensive – written communication makes use of relatively more resources as compared to oral communication. Moreover, for short information transmission to limited people at short distances, written mode may prove to be very expensive.

ii. Time consuming – Written communication eats up a lot of time both at the time of drafting the message and for it transmission. On receiving, if there are some clarifications and queries they are again written back which are answered back and so on causing a delayed communication process.

iii. Limited applicability – Written communication has a limited applicability for those who are literate. For people who cannot read and write, possibility of written mode of communication is totally ruled out.

iv. Inflexibility – Written documents once transmitted cannot be amended and altered. Even for the slightest mistakes. Either it has to be redrafted or kept unaltered making it inflexible.

v. Lack of personal touch – Written mode of communication is very formal and impersonal in nature. Though written mode, one cannot express the intensity of feelings which otherwise would have been possible through words or gestures.

vi. Lack of secrecy – Written documents being in black and white and stored for future reference, are always subject to the possibility of being leaked or stolen by somebody.

vii. Delay in feedback – Unlike oral communication, where queries, clarifications and comments move instantaneously from receiver. In written communication, there is a time gap between sending and receiving of message. Moreover, feedback and response also being in written form again takes time and cause delay.

B. Non-verbal communication

It is not necessary that communication is possible only through verbal means. Communication takes place without words, spoken or written. Many of research study has established the efficacy of non-verbal communication. It finds expression through gestures, postures, body movement and physical appearance. Message are expressed through non-linguistic means. When a person refuses to speak or write, he can communicate non-verbally.

Non-verbal communication may be defined as medium of transmitting information without the use of words. In other words, it encompasses all such forms of communicating which we cannot read or write. Normally, non – verbal communication is used to enhance the effectiveness of oral communication. This method of communication is very helpful in expressing feelings, concern, emotions, gratitude and also where there is absence of common language.

Non-verbal communication is expressed by various ways:

1. Gestures
2. Facial expressions
3. Touch
4. Attire
5. Body posture

Advantages

i. Easy presentation: Information can be easily transmitted through non-verbal mode as no expertise skill is required in preparation and at times, it is extempore such as – a handshake or pat on the back.

ii. Good substitution: Non-verbal communication may prove to be a very useful substitution for oral or written communication is situations where either verbal communication is not possible or where it is disturbed or blocked through noise or interruption. For example, in examination hall, asking time by putting finger on wrist or nodding for saying yes.

iii. Combats linguistic barrier: Wherever sender and receiver of information do not share a common language, non-verbal communication is the most effective communication. It enables to communicate through gestures and symbols.

iv. Complements oral communication: Certain gestures and body postures, when used, adds on to the meaning of oral communication. With non-verbal communication, oral communication may become more assertive and impactful. For example, while appreciating an employee, complementing it with clapping will create greater impact.

v. Useful for illiterate peoples: Use of gestures, facial expressions, eye contact and proximity, touching, etc., without using any spoken or written words to communicate is very much helpful for illiterate people.

vi. Instant impact: No-verbal expressions have an immediate impact as the receivers perceive them quickly and instantly. For example, showing a flag to make people gather at a place is more effective than shouting. This quality of being speedy in conveyance and response makes non-verbal methods extremely useful and imperative in many situations. For example, ringing of a bell at different time slots indicating change of class in colleges or schools has no verbal alternative.

vii. Indicator of belongingness: Non-verbal communication may also be used to assign or mark the identity of some people or they being belonging to a particular group. For example, wearing a cap to show belongingness to a particular party, wearing a dress of particular colour, air hostesses have a specific attire, etc.

Disadvantages

i. Differences in non-verbal cues across culture: Non-verbal cues are subject to different interpretations in a multicultural set up. For example, in a particular culture wearing while may be considered as suspicious whereas in another it may be an indication of grief. These differences lead to misunderstanding and confusion.

ii. Loses effect after occurrence: No doubt, non –verbal communication has a very strong instant and speedy effect, but it lacks the power of retention as it cannot be read or heard repeatedly. In some case, it cannot produce any effect at all. For example, a gesture may go unnoticed.

iii. Absence of permanent record: Non-verbal communication cannot be retained as a permanent record. Being a non word based communication. This mode is subject to be forgotten very easily on the grounds of "out of sight, out of mind" and thus, is not authentic method.

iv. Possibility of distortion of message: Possibility of distortion of message in non-verbal communication is very high because non-verbal expressions varies across countries, cultures, religions and even varies from person to person. Thus, the meaning of a message may be wrongly interpreted leading to ineffective communication.

v. Vague and imprecise: There is no ready reckoner or a dictionary which would help to decode the meaning of non-verbal cues and give them some interpretation. This makes non-verbal communication quite vague and imprecise.

(C) Body language/Gestural communication

Gestural communication fits with both verbal and written communication. Gestures are used in verbal communication because we instinctively feel that words do not – indeed cannot fully express our ideas, thoughts, feeling and emotions. So our hands and face, in fact the entire body automatically jumps in to add meaning to the communication. In a written communication, too, the sender writing the receiver's name, or concluding the letter by writing 'with regards' will give an intimate personal touch to the communication.

A gestural communication is a non-verbal, non-vocal communication in which the communicator makes visible bodily actions to transmit a message in place of communicating it by word of mouth. Sometimes, a gestural communication may accompany a verbal communication.

Gestural communication is admission of the fact that words of any language cannot fully express all the ideas, thoughts, feelings and emotions. Gestures, whether by way of wave of hand, use of fingers, facial expression- rolling, shrinking or widening of eyes, pouting of lips, inflation or shrinking of nose, rolling of tongue, *etc.* – are used to give added meaning and effect to the communication.

- Although the most common methods of communication are carried out orally or in writing, when it comes to management techniques, the power of non-verbal communication must never be underestimated.
- Smile, gestures and several other body movements send out a message to the people around you. You need to be mindful of this while dealing with your employees and customers.
- Always remember to maintain eye contact. This would show that you are serious and confident about what is being said.

Gestures may assume many forms. The dress one wears, the scent one uses, expression on the face (smile, scowl, sneer, smirk), the food prepared by spouse – all convey feelings, emotions, ideas or thoughts. Observes a politician addressing a meeting.

i. Communication through facial gestures:

Facial expression, like other bodily gestures, communicates a lot. A nod will indicate understanding, a frown or a wrinkle of the nose suggests a feeling of disgust or inability to understand the message. Emotions expressed through face convey many things – happiness, sadness, fear, anger, surprise, contempt, disgust, etc. As far other gestures, rubbing of nose conveys rejection or doubt or an indication that the communicator is lying. Rubbing of eyes suggests doubt or disbelief. If the receiver has his head resting in his hands and with downcast eyes. It means expression of boredom and lack of interest in the message.

ii. Communication through body postures

There are certain body postures that, according to psychologists, suggest specific feelings and emotions. Thus, standing erect indicates self confidence and standing with hands on hips indicates readiness to act. Sitting cross legged indicates disinterestedness or unwillingness. Sitting with legs open indicates the person is relaxed and open to ideas. Arms crossed on the chest suggest a defensive posture; walking with hands thrust in pockets suggests sadness and misery. Hands covering indicate the person is thinking and evaluating the communication.

Hands clasped behind the back convey anger, frustration and apprehension. Rubbing hands suggests the person in anticipating something. Tapping and drumming fingers or rolling paper weight on the table indicates impatience. Patting or fondling of hair conveys lack of confidence and nail biting – a feeling of insecurity or nervousness.

Gestural communication also adds effect to a written communication. In a typed letter let the manager write the name of his subordinate in his own hand and see how happy the subordinate feels.

Broadly there are four main types of communication we use on a daily basis: Verbal, nonverbal, written and visual.

A. Verbal Communication

Verbal communication involves the usage of words while the delivery of the intended message is being made. It can be one-on-one, over the phone or in group settings, etc. it is a methods of effective communication that is personal and has to be used more than only phone calls and emails whenever possible. The medium of the message in case of verbal communication is oral. Simple speaking is verbal communication. Seeing the person with whom face-to-face communication is taking place helps in gauging the response of that person by understanding their body language and also assists in active participation of the dialogue. Thus, written as well as oral communication are the two main types of verbal communication.

1. Storytelling

Storytelling is a form of verbal communication that is effective. A vital organisational function is served by it as it helps in the construction of common meanings for the individuals in the organisation. Stories assist in clarifying key values and also help in demonstrating the way thing are done in an organisation. The tone, strength and frequency of the story have a relation with higher organisational commitment. The quality of stories that are told by the entrepreneurs is related to the capability of securing capital for their organisations. Stories can help in serving to reinforce as well as perpetuate the culture of an organisation.

2. Crucial conversations

Even though the process is similar, more skill, reflection and planning is needed for high-stakes communications in comparison to daily interactions at work. Examples of communication that involves high-stakes consist of presenting the venture capitalist a business plan and asking for a raise. Apart from these, at several times in our professional lives, we have conversations that are crucial – discussions in which stakes are high, emotions run strong and opinions may differ.

Verbal communication is the use of language to transfer information through speaking or sign language. It is one of the most common types, often used during presentations, video conferences and phone calls, meetings and one-on-one conversations. Verbal communication is important because it is efficient. It can be helpful to support verbal communication with both nonverbal and written communication.

There are a few steps to develop verbal communication skills:

- **Use a strong, confident speaking voice.** Especially when presenting information to a few or a group of people, be sure to use a strong voice so that everyone can easily hear you. Be confident when speaking so that ideas are clear and easy for others to understand.

- **Use active listening.** The other side of using verbal communication is intently listening to and hearing others. Active listening skills are key when conducting a meeting, presentation or even when participating in a one-on-one conversation. Doing will help to grow as a communicator.
- **Avoid filler words.** It can be tempting, especially during a presentation, to use filler words such as "um," "like," "so" or "yeah." While it might feel natural after completing a sentence or pausing to collect thoughts, it can also be distracting for audience. Try presenting to a trusted friend or colleague who can call attention to the times you use filler words.

B. Nonverbal methods of Communication

The most important thing in communication is hearing what isn't said.

(i) Peter Drucker

Communication by sending as well as receiving wordless messages is a part of nonverbal communication. Such messages are a vital part of any communication and generally provide reinforcement to verbal communication. However, they can also convey their own feelings and thoughts.

(ii) Physical nonverbal communication

A major role is played by the nonverbal communication in the way to communicate with other people. It cannot be relied as a sole means of communication except in cases where the person makes use of sign language. Physical nonverbal communication or the body language consists of body posture, facial expressions, eye contact, gestures like a pointed finger, wave and the like, touch, tone of voice, overall movements of the body and others.

The most common among all the nonverbal communication are the facial expressions. For example, a frown or a smile conveys emotions that are distinct and are difficult to express by means of verbal communication. Research has shown that nonverbal cues or body language along with facial expressions, tone of voice and body stance account for almost 55% of all the communication that takes place. As per the research, only 7% of the message's comprehension by the receiver is based on the actual words of the sender, 38% is on the basis of para language (the volume, tone and pace of speech) while 55% is on the basis of nonverbal cues (body language).

Research has also revealed that nonverbal cues may also have an effect on whether you are successful in getting a job offer. Judges who examine the videotapes of the applicants were able to make assessments of their social skills even when the sound was off. They had a look at the gesturing rate, time that was spent on talking and the dress formality for determining the candidates who would be most successful for the job socially. Thus, it is essential to consider the way one appears in the business and also what one says. Our emotions are conveyed by our facial muscles. A silent message can be sent without uttering a word. A change in our emotional state is visible in the change of our facial expression.

For example, if we focus on being confident prior to an interview, the confidence will be conveyed to the interviewer by our face. Putting up a smile even when feeling stressed can reduce the stress levels of the body.

(iii) Paralanguage

A significant component of nonverbal communication is the way something is said, instead of what is said actually. This consists of style of speaking, tone, emotion, stress, pitch, intonation and voice quality. It helps in the communication of interest, approval or the absence of it. According to research, 38% of all communications are accounted for by the tone of the voice.

The other nonverbal communication forms generally communicate the personality of a person. These consist of –

- **Aesthetic communication** or creative expressions like painting, dancing, etc.
- **Appearance** or the style of grooming and dressing that communicates the personality of a person.
- **Space language** like landscapes and paintings communicate taste as well as social status.
- **Symbols** like ego-building, status and religious symbols.

For being effective communicators, it is important to align our tone, appearance and body language with the word that are being conveyed by us. Research has indicated that when individuals lie, they blink more often, shrug and shift their weight. Tone is another element of communication that is nonverbal. A tone that is different can alter the message's perceived meaning. The emphasis on a particular word illustrates the way a tone can convey a lot of information.

Nonverbal communication is the use of body language, gestures and facial expressions to convey information to others. It can be used both intentionally and unintentionally. For example, you might smile unintentionally when you hear a pleasing or enjoyable idea or piece of information. Nonverbal communication is helpful when trying to understand others' thoughts and feelings.

If they are displaying "closed" body language, such as crossed arms or hunched shoulders, they might be feeling anxious, angry or nervous. If they are displaying "open" body language with both feet on the floor and arms by their side or on the table, they are likely feeling positive and open to information.

There are a few steps to develop nonverbal communication skills:

- **Notice how your emotions feel physically.** Throughout the day, as you experience a range of emotions (anything from energized, bored, happy or frustrated), try to identify where you feel that emotion within your body. For example, if you're feeling anxious, you might notice that your stomach feels tight. Developing self-awareness around how your emotions affect your body can give you greater mastery over your external presentation.
- **Be intentional about your nonverbal communications.** Make an effort to display positive body language when you feel alert, open and positive about your surroundings.

You can also use body language to support your verbal communication if you feel confused or anxious about information, like using a furrowed brow. Use body language alongside verbal communication such as asking follow up questions or pulling the presenter aside to give feedback.

- **Mimic nonverbal communications you find effective**. If you find certain facial expressions or body language beneficial to a certain setting, use it as a guide when improving your own nonverbal communications. For example, if you see that when someone nods their head it communicates approval and positive feedback efficiently, use it in your next meeting when you have the same feelings.

C. Written Communication

Written communication cannot be avoided in the workplace. It is present everywhere and consists of traditional paper and pen documents and letters, text chats, emails, typed electronic documents, reports, SMS and anything else that might be conveyed by the use of written symbols like language which includes any documents that are a part of the day to day business life. Such methods of communication is indispensable for any formal business communication and also for the issue of legal instructions.

The methods of communication that mainly make use of written communication consist of formal business proposals, press releases, memos, contracts, brochures, handbooks and the like. The effectiveness of written communication is will depend on the style of writing, vocabulary, grammar along with clarity. Written communication is most suitable in cases that require detailed instructions, when something has to be documented or in situations where the person is too far away to speak in person or over the phone. Some of the written methods of communication include –

1. Social Media

Tools of the social media such as facebook can be used in an effective way in the organisations for the purpose of communicating between locations, divisions, departments and employees. Certain tools such as Yammer are specifically designed for creating social media networks in the confines of a particular domain of business.

2. Email

The email communication in organisations has become very common and is a highly essential tool for sharing of information with one, hundreds as well as thousands of employees. It is usually a significant component of the communication plan and even though it may not be a replacement for face-to-face or other methods of communication, it is easily accessible and inexpensive.

3. Oral Communication

The other type of verbal communication includes the spoken word, either over the phone or face-to-face, video conferencing, voice chat or any other medium. Different types of informal communications like the informal rumour mill or grapevine and formal communications such as conferences and lectures are the forms of effective oral communication. The use of oral communications is made in discussions as well as conversations that are informal and causal. How effective the oral communication is depend on the speed, volume, pitch, voice modulation, clarity of speech and also the non-verbal communications like visual cues and body language.

Some of the important oral methods of communication are –

(i) Face-to-Face Communication

This is usually preferred methods of communication although it is not realistic all the time, especially in organisations that are based in several locations around the globe. However, tools like videoconferencing make approximate face-to-face communication possible and help even the large organisations in creating personal connections between the management and the staff.

(ii) Meetings

Meetings are common in almost all the business settings and now it has become possible to augment them via use of the tools of technology which allow the participants in the remote locations to participate in the discussions, although they are not present physically on the meeting site.

While using verbal communication, the person needs to be aware of his tone of voice, inflection as well as speed. Use of sarcasm and angry tones should be avoided otherwise the person with whom you are communicating will go on the defensive. While speaking to a team or to a group of people, you should ensure that you speak loudly as well as clearly so that everyone is able to understand what you are saying. Use of the microphone should be made if a group in is being addressed in a large room. Verbal communication is best used when something has to be discussed in detail or when someone has to be complimented or reprimanded. Verbal communication makes the conveying of thoughts faster and easier and is the most successful methods of communication. However, it makes up just 7% of all the human communication.

Written communication is the act of writing, typing or printing symbols like letters and numbers to convey information. It is helpful because it provides a record of information for reference. Writing is commonly used to share information through books, pamphlets, blogs, letters, memos and more. Emails and chats are a common form of written communication in the workplace.

There are a few steps to develop written communication skills:

- **Strive for simplicity.** Written communications should be as simple and clear as possible. While it might be helpful to include lots of detail in instructional communications, for example, you should look for areas where you can write as clearly as possible for your audience to understand.
- **Don't rely on tone.** Because you do not have the nuance of verbal and nonverbal communications, be careful when you are trying to communicate a certain tone when writing. For example, attempting to communicate a joke, sarcasm or excitement might be translated differently depending on the audience. Instead, try to keep your writing as simple and plain as possible and follow up with verbal communications where you can add more personality.

- **Take time to review your written communications.** Setting time aside to re-read your emails, letters or memos can help you identify mistakes or opportunities to say something differently. For important communications or those that will be sent to a large number of people, it might be helpful to have a trusted colleague review it as well.
- **Keep a file of writing you find effective or enjoyable.** If you receive a certain pamphlet, email or memo that you find particularly helpful or interesting, save it for reference when writing your own communications. Incorporating methods or styles you like can help you to improve over time.

D. Visual Communication

Visual communication is the fourth methods of communication which takes place through the help of visual aids like colour, illustration, graphic design, drawing, typography, signs and other electronic resources.

Visual communication comprising charts and graphs generally reinforces written methods of communication and in majority of the situations replaces written communication completely. "A picture is worth a thousand words" is an adage which shows that visual communication can on several occasions be more powerful than verbal as well as nonverbal communication. The developments in technology have made expression of visual communication easier than what it was before.

Visual communication is the act of using photographs, art, drawings, sketches, charts and graphs to convey information. Visuals are often used as an aid during presentations to provide helpful context alongside written and/or verbal communication. Because people have different learning styles, visual communication might be more helpful for some to consume ideas and information.

There are a few steps to develop visual communication skills:

- **Ask others before including visuals.** If you are considering sharing a visual aid in your presentation or email, consider asking others for feedback. Adding visuals can sometimes make concepts confusing or muddled. Getting a third-party perspective can help you decide whether the visual adds value to your communications.
- **Consider your audience.** Be sure to include visuals that are easily understood by your audience. For example, if you are displaying a chart with unfamiliar data, be sure to take time and explain what is happening in the visual and how it relates to what you are saying. You should never use sensitive, offensive, violent or graphic visuals in any form.

To make improvements to your communication skills, set personal goals to work through the things you want to accomplish step by step. It might be helpful to consult with trusted colleagues, managers or mentors to identify which areas would be best to focus on first. There is no one correct way of communicating but you need to have awareness about what is appropriate and in which situation. The chosen methods of communication depends on the situation.

■■■■

Chapter - 8 Communication Process

- Communications is fundamental to the existence and survival of humans as well as to an organization. It is a process of creating and sharing ideas, information, views, facts, feelings, etc. among the people to reach a common understanding.
- Communication is the key to the Directing function of management.
- Communication refers to the process by which the information is transmitted and understood between two or more people. Transmitting the sender's intended meaning is the essence of effective communication.
- Communication involves two people – a sender and a receiver.
- One person alone cannot communicate.
- According to Rogers and Shoemaker (1971) – communication is the process by which messages are transferred from a source to receiver.
- According to Leagans (1961) – communication is the process by which two or more people exchange ideas, facts, feelings or impressions in ways that each gains a common understanding of the meaning, intent and use of messages.

Concept of communication process:

Each act of communication has atleast three phases – expression, interpretation and response. If the expression is not clear, the interpretation accurate and the response logical one's effort to communicate will not succeed. As it is difficult to control how an audience shall interpret the message and respond to it, a powerful communication effort by the extension agent must be constantly extended.

Types of Communication

1. Formal Communication

Formal communications are the one which flows through the official channels designed in the organizational chart. It may take place between a superior and a subordinate, a subordinate and a superior or among the same cadre employees or managers. These communications can be oral or in writing and are generally recorded and filed in the office.

Formal communication may be further classified as vertical communication and horizontal communication.

i. Vertical Communication

Vertical Communications as the name suggests flows vertically upwards or downwards through formal channels. Upward communication refers to the flow of communication from a subordinate to a superior whereas downward communication flows from a superior to a subordinate.

Application for grant of leave, submission of a progress report, request for loans etc. are some of the examples of upward communication. Sending notice to employees to attend a meeting, delegating work to the subordinates, informing them about the company policies, etc. are some examples of downward communication.

ii. Horizontal Communication

Horizontal or lateral communication takes place between one division and another. For example, a production manager may contact the finance manager to discuss the delivery of raw material or its purchase.

Types of communication networks in formal communication

Single chain: In this type of network communications flows from every superior to his subordinate through a single chain.

Wheel: In this network, all subordinates under one superior communicate through him only. They are not allowed to talk among themselves.

Circular: In this type of network, the communication moves in a circle. Each person is able to communicate with his adjoining two persons only.

Free flow: In this network, each person can communicate with any other person freely. There is no restriction.

Inverted V: In this type of network, a subordinate is allowed to communicate with his immediate superior as well as his superior's superior also. However, in the latter case, only ordained communication takes place.

2. Informal Communication

Any communication that takes place without following the formal channels of communication is said to be informal communication. The Informal communication is often referred to as the 'grapevine' as it spreads throughout the organization and in all directions without any regard to the levels of authority.

The informal communication spreads rapidly, often gets distorted and it is very difficult to detect the source of such communication. It also leads to rumors which are not true. People's behaviour is often affected by the rumors and informal discussions which sometimes may hamper the work environment.

Communications Process

- The process of communication refers to the transmission or passage of information or message from the sender through a selected channel to the receiver overcoming barriers that affect its pace.

- The process of communication is a cyclic one as it begins with the sender and ends with the sender in the form of feedback. It takes place upward, downward and laterally throughout the organization.
- The process of communication as such must be a continuous and dynamic interaction, both affecting and being affected by many variables.
- Communication process consists of certain steps where each step constitutes the essential of an effective communication.

The following is a brief analysis of the important steps of the process of communication.

1. Sender: The communication process is laid by the person who transmits or sends the message. He is the sender of the message which may be a thought, idea, a picture, symbol, report or an order and postures and gestures, even a momentary smile. The sender is therefore the initiator of the message that need to be transmitted. After having generated the idea, information etc. the sender encodes it in such a manner that can be well-understood by the receiver.

2. Message: Message is referred to as the information conveyed by words as in speech and write-ups, signs, pictures or symbols depending upon the situation and the nature and importance of information desired to be sent. Message is the heart of communication. It is the content the sender wants to convey to the receiver. It can be verbal both written and spoken; or non-verbal *i.e.* pictorial or symbolic, etc.

3. Encoding: Encoding is putting the targeted message into appropriate medium which may be verbal or non-verbal depending upon the situation, time, space and nature of the message to be sent. The sender puts the message into a series of symbols, pictures or words which will be communicated to the intended receiver. Encoding is an important step in the communication process as wrong and inappropriate encoding may defeat the true intent of the communication process.

4. Channel: Channel(s) refers to the way or mode the message flows or is transmitted through. The message is transmitted over a channel that links the sender with the receiver. The message may be oral or written and it may be transmitted through a memorandum, a computer, telephone, cell phone, apps or televisions.

Since each channel has its advantages and disadvantages, the choice of proper selection of the channel is paramount for effective communication.

5. Receiver: Receiver is the person or group who the message is meant for. He may be a listener, a reader or a viewer. Any negligence on the part of the receiver may make the communication ineffective. The receiver needs to comprehend the message sent in the best possible manner such that the true intent of the communication is attained. The extent to which the receiver decodes the message depends on his/her knowledge of the subject matter of the message, experience, trust and relationship with the sender.

The receiver is as significant a factor in communication process as the sender is. It is the other end of the process. The receiver should be in fit condition to receive the message, that is, he/she should have channel of communication active and should not be preoccupied with other thoughts that might cause him/her to pay insufficient attention to the message.

6. Decoding: Decoding refers to interpreting or converting the sent message into intelligible language. It simply means comprehending the message. The receiver after receiving the message interprets it and tries to understand it in the best possible manner.

7. Feedback: Feedback is the ultimate aspect of communication process. It refers to the response of the receiver as to the message sent to him/her by the sender. Feedback is necessary to ensure that the message has been effectively encoded, sent, decoded and comprehended.

It is the final step of the communication process and establishes that the receiver has received the message in its letter and spirit. In other words, the receiver has correctly interpreted the message as it was intended by the sender. It is instrumental to make communication effective and purposeful.

Consider the following points related to the feedback involved in the process of communication –

- It enhances the effectiveness of the communication as it permits the sender to know the efficacy of his message.
- It enables the sender to know if his/her message has been properly comprehended.
- The analysis of feedbacks helps improve future messages. Feedback, like the message, can be verbal or nonverbal and transmitted through carefully chosen channel of communication.
- We can represent the above steps in a model as the model of communication process.

Types of Feedback

There are four types of feedback. The types are as follows –

- Negative Feedback or corrective comments about past behaviour
- Positive Feedback or affirming comments about future behaviour
- Negative *feedforward* or corrective comments about future behaviour
- Positive *feedforward* or affirming comments about future behaviour

8. Noise

It refers to any obstruction that is caused by the sender, message or receiver during the process of communication. For example, bad telephone connection, faulty encoding, faulty decoding, inattentive receiver, poor understanding of message due to prejudice or inappropriate gestures, etc.

The model of communication process

The model of communication process are as :

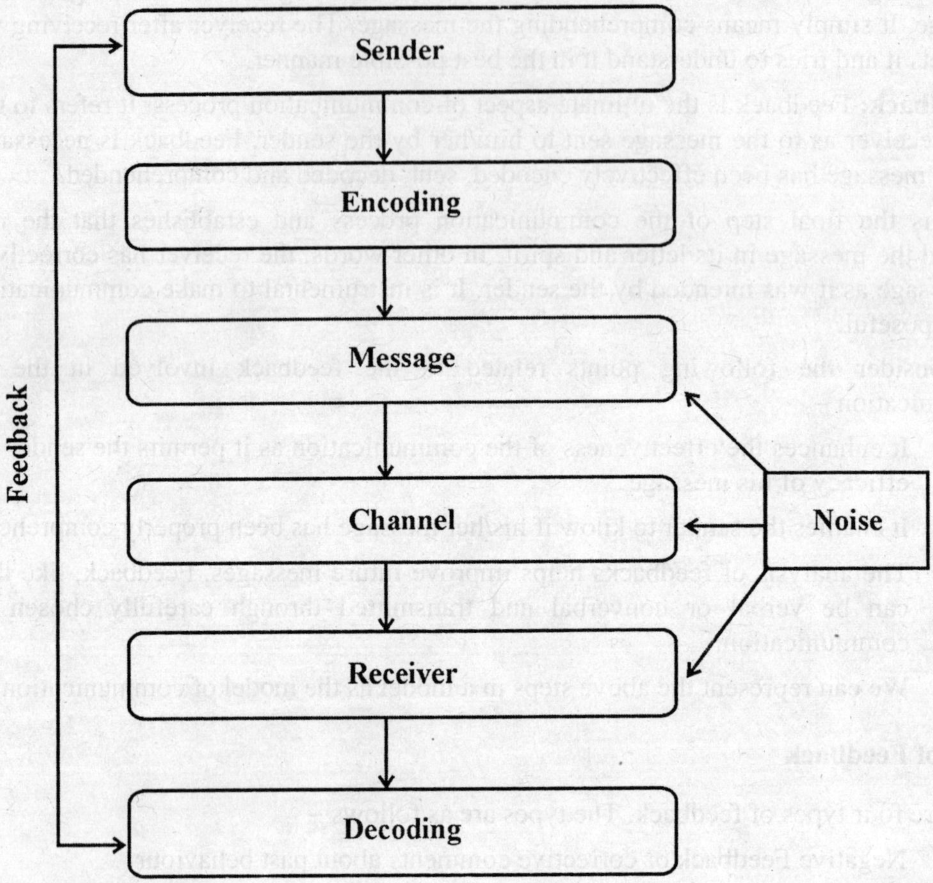

Importance of Communication

1. The basis of co-ordination: The manager explains to the employees the organizational goals, modes of their achievement and also the interpersonal relationships amongst them. This provides coordination between various employees and also departments. Thus, communications act as a basis for coordination in the organization.

2. Fluent working: A manager coordinates the human and physical elements of an organization to run it smoothly and efficiently. This coordination is not possible without proper communication.

3. The basis of decision making: Proper communication provides information to the manager that is useful for decision making. No decisions could be taken in the absence of information. Thus, communication is the basis for taking the right decisions.

4. Increases managerial efficiency: The manager conveys the targets and issues instructions and allocates jobs to the subordinates. All of these aspects involve communication. Thus, communication is essential for the quick and effective performance of the managers and the entire organization.

5. Increases cooperation and organizational peace: The two-way communication process promotes co-operation and mutual understanding amongst the workers and also between them and the management. This leads to less friction and thus leads to industrial peace in the factory and efficient operations.

6. Boosts morale of the employees: Good communication helps the workers to adjust to the physical and social aspect of work. It also improves good human relations in the industry. An efficient system of communication enables the management to motivate, influence and satisfy the subordinates which in turn boosts their morale and keeps them motivated.

Barriers to communication

The communication barriers may prevent communication or carry incorrect meaning due to which misunderstandings may be created. Therefore, it is essential for a manager to identify such barriers and take appropriate measures to overcome them. The barriers to communication in organizations can be broadly grouped as follows:

1. Semantic Barriers: These are concerned with the problems and obstructions in the process of encoding and decoding of a message into words or impressions. Normally, such barriers result due to use of wrong words, faulty translations, different interpretations, etc.

For example, a manager has to communicate with workers who have no knowledge of the english language and on the other side, he is not well conversant with the hindi language. The language is a barrier to communication as the manager may not be able to communicate properly with the workers.

2. Psychological barriers: Emotional or psychological factors also act as barriers to communication. The state of mind of both sender and receiver of communication reflects in effective communication. A worried person cannot communicate properly and an angry recipient cannot understand the message properly.

Thus, at the time of communication, both the sender and the receiver need to be psychologically sound. Also, they should trust each other. If they do not believe each other, they cannot understand each other's message in its original sense.

3. Organizational barriers: The factors related to organizational structure, rules and regulations authority relationships, etc. may sometimes act as barriers to effective communication. In an organization with a highly centralized pattern, people may not be encouraged to have free communication. Also, rigid rules and regulations and cumbersome procedures may also become a hurdle to communication.

4. Personal barriers: The personal factors of both sender and receiver may act as a barrier to effective communication. If a superior thinks that a particular communication may adversely affect his authority, he may suppress such communication.

■■■■

Chapter - 9 **Audio-Visual Aids**

Audio-Visual Aids

Audio-visual aids are instructional devices which are used to communicate message more effectively through sound and visuals. These may be used for literate as well as illiterate people.

Audio aids: The instructional device through which the message can only be heard are known as Audio-aids.

Visual aids: The instructional device through which the message can only be seen are known as visual aids.

Audio-visual aids: The instructional device through which the message can be heard and seen simultaneously are known as audio-visual aids.

The use of audio-visual aids has the following advantages:

1. Capture audience attention an arouse their interest.
2. Highlight main points of the message clearly.
3. The possibility of misinterpreting concepts is reduced.
4. Structure the learning process more effectively.
5. Message perceived with several senses are understood and retained better.
6. Provide experience otherwise difficult to get.
7. Help reach more people irrespective of their level of literacy and language.
8. Speed up the process of learning.
9. Can adopt the speed of presentation to the reactions of the audience.
10. Save time of teacher and learner.

Classification of audio-visual aids

The audio-visual aids may be classified into three categories as follows:

A. Audio-aids	**B. Visual –aids**	
	Non projected	**Projected**
1. Tape –recorder	1. Chalk board	1. Slides
2. Public address system	2. Bulletin board	2. Film strip

3. Telephone	3. Picture and photograph	3. Opaque projection
	4. Flannel graph, flash card, Flip chart	4. Overhead projector
	5. Poster	
	6. Diagram, map, chart and graph	
	7. Specimen, model, diorama	
	8. Translide	

C. Audio-visual aids

Projected	**Non projected**
1. Dramma	1. Motion picture (cinema)
2. Puppet show	2. Video
3. Talking doll	3. Folk media
	4. Smart phone/mobile
	5. Webinars

Audio-visual aids may also be classified as

1. **Display type:** Display type of visuals are those which are spread before the audience for viewing, who get the message by looking at them. These are poster, bulletin, board, models, exhibits etc.
2. **Presentation type:** Presentation type visuals are presented or projected before the audience for viewing, but at the same time one explains or presents the message of the visuals, so that the audience gets a meaningful understanding of them. These are flash card, slides, filmstrip etc.

Choice of audio-visual aids

Audio-visual aids are used singly or in combination with other aids and extension methods. The choice of audio-visual aids shall depend on a number of criteria.

1. **Teaching objectives:** whether to give information, to impart skill or to bring change in attitude etc.
2. **Nature of subject matter being taught:** particular aspect of the technology and whether understanding depends on seeing or not.
3. **Nature of audience:** their age, education, interest, experience, knowledge, intelligence etc.
4. **Size of the audience:** small or large
5. Availability of equipments, materials and funds

6. Skill and experience of the extension agent in preparation and use of audio-visual aids.

A. Audio-aids

1. Tape recorder: Tape recorder is an equipment for recording sound on magnetic tape by electro-magnetic process, which may be played back when needed. The tape may be celluloid, plastic or high tensile polyester film. Tape recorder is suitable for use in meeting, training programme, campaign, recording radio programme. A tape recorder for extension work should have the following features:

- Simple with minimum number of buttons and knobs.
- Light weight and easy to carry.
- Battery operated can also be used on AC/DC power battery.
- Variable speed with 1 to 2 recoding tracks.
- Sturdy in construction and moderate in cost.

2. Public address system: Public address system is a set of equipments to amplify sound so that it is audiable to a large audience over a distance. It is useful for extension programmes involving a large number of people such as mass meeting, training programme, field day or farmers day, campaign, exhibition etc. The public address system has three components *viz.,* microphone, amplifier and loudspeaker. The microphone is connected to the input terminal of the amplifier and the loudspeaker to the output terminal. It may be run on dry/wet battery or AC/DC power supply.

3. Telephone: Telephone is a system of equipments through which people can communicate both ways to distant places. Only two people can communicate at a time through a telephone, the system serves many people in a given areas with the establishment of telephone in each gram panchayat office at the village level, rural telephone has become a reality in India.

B. Visual aids

(i). Non projected visual aids

1. Chalkboard: Chalkboard is a board for writing with chalk. Chalkboard is the simplest, cheapest, most convenient and widely used visual aids. Wooden board, coloured with black is mostly used, hence it is also known as black board. However, ground glass, roll up materials, coloured in black, olive green or blue may be used. Chalk board is suitable for use in lecture, training programme, group meeting etc.

2. Bulletin board: Bulletin board is a board for display of message. Bulletin board may be made of soft insulation board or perforated masomite board and may or may not be covered with glass. Paper containing the message may be fixed by board pin on the soft insulation board. The message on the bulletin board may be in the form of words, graphs, charts, photographs, illustrations, publications etc. and may be used in the communication centre and various types of extension programme.

3. Picture and photograph: Picture is a representation made by drawing, painting or photography which gives accurate idea of an object. A good picture may tell a story without using a single word. Colour pictures and blow-up photographs have more appeal. Pictures and photographs are used in various ways in extension work such as training programme, farm publication, campaign, exhibition, slide, filmstrip, motion picture, television, newspaper, display etc.

4. Flannel graph, flash card, film chart: Flannel graph, flash card and flip chart are visual aids in which the message are written or drawn on thick paper and presented step-by-step by the extension agent to the audience to synchronize with the talk. This produces a dramatic effect on the audience. These are generally used in class room situation, rather than in field extension work.

5. Poster: Poster is a placard displayed in a public place with the purpose of creating awareness amongst the people. A poster is generally seen from a distance and the person glancing at it seldom has the time or inclination to stop and read.

A good poster should have the following properties:

i. It must be able to attract attention.

ii. It must convey the message quickly.

iii. It must lead to action.

Hand drawn posters may be used in training programme, group meeting, farmer's day etc. while, printed posters may be used in large numbers of campaign, exhibition etc. Wall writing and sign boards may be made by following the principles of poster and are effective for communication.

6. Diagram, map, chart and graph: Diagram, map, chart and graph are visuals where information is summarized and presented in a more or less abstract form. Diagram is line drawing of an object or an idea, a map is an informative diagram of an area, a chart contains information in tabular form and a graph is a diagrammatic representation of the relationship between variables. Diagram, map, chart and graph are used in bulletin board, method demonstration, group meeting, training programme, farm publication, campaign, exhibition, newspaper , television etc.

7. Specimen, model, Diorama

Specimen is a sample which represents the whole.

Model is miniature replica of an object.

Diorama is a scenic representation of the original, with specimen, model and painting.

Specimen and model are used in method demonstration, group meeting, training programme, exhibition etc. Diorama is greatly used in exhibition and in communication centres.

8. Translide: Translide are transparent big size photographs which are displayed by providing light at the back. Translides are costly, but produce beautiful life like visuals. These may be used in communication centre, information centre and exhibition.

(ii) Projected visual – aids

1. Slides: Slide is a transparent mounted picture which is projected by focusing light through it. The projection may be made on roll-back screen or on white wall. Slides of 35 mm film mounted on individual cardboard frame are more common and are extensively used in extension programme. Glass slides are generally used in cinema halls. Slides are used in training programme, seminar, workshop, group meeting, campaign, exhibition, television etc.

2. Film strip: Film strip is a continuous strip of film consisting of a small number of individual frames or pictures arranged in a sequence. Each strip may contain about 30 to 60 frames of photographs, diagrams, drawings or lettering. The pictures may have explanatory titles. Film strip are generally used in training programmes.

3. Opaque projection: In opaque projection a large apertured projector is used to project opaque materials. Opaque projector is also known as epidioscope. Pictures, drawings, diagrams directly from books, magazines etc or drawn on sheets of paper can be projected. The size of projectable material should be around 25 cm x 25 cm. Opaque projection may be used in training programme, group meeting etc.

4. Overhead projection: In overhead projection, the picture is projected over the head of the speaker on the screen. This is accomplished by an overhead projector. Drawing, diagram, lettering etc. are made on transparent sheets and are put on the glass plateform of the overhead projector, through which strong light is passed. Overhead projection is used in training programme, group meeting, seminar, symposium, workshop etc. Printing on transparencies may be made through photographic, Xerox electronic processes. The transparencies may be in colour or in black and white, and may be repeatedly used.

C. Audio-visual aids

(i) Projected audio-visual aids

Drama, Puppet show, Talking doll

Drama is a theatrical performance around a theme by some people who have rehearsed for it.

In puppet show the dolls are made to perform by controlled movement of the puppeter from outside.

Both the media requires adequate preparation and rehearsing. Puppet show may be held in daylight or at night, drama is generally performed at a night.

Talking doll is an innovation and may be used to create general awareness of a group of peoples about an extension programme.

Dramma, puppet show and talking doll provide dramatic experience to the audience and may be used to communicate messages on broad social themes pertaining to rural life, rather than transfer of technology programme.

(ii) Non projected audio-visual aids

1. Motion picture (Cinema): Motion picture is the projected image of a positive print of a sound film taken by a movie camera, which produces synchronized movement of picture and sound. The film may be in black and white or in colour. Colour film have more appeal. The sound film may be of 16 mm, 35 mm and 70 mm width. The film with 16 mm width is generally used in extension work.

Motion picture is used in training programme, group meeting, method demonstration, campaign, exhibition, mass meeting, television etc.

2. Video: Video is an electronic audio-visual system for preparing video tapes of programmes and events, which could be projected through a receiving set with moving images and synchronized sound. The complementarily of video, motion, picture and television and their differences shall help in understanding the special features and potential of video for use in extension.

3. Folk media: Folk media is a communication system embedded in the culture which existed before the arrival of mass media and still exists as a vital mode of communication in many parts of the world, presenting a certain degree of continuity, despite changes. To a society, folk media are culturally more compatible.

The folk media consist of a variety of forms – folk theatre, puppetry, story-telling, folk painting, folk dance, folk music, ballad and mime. The traditional uses of folk media were primarily for entertainment, mutual participation at the social level, religious activity etc. The folk media may be used to increase the effectiveness of communication and bring about greater participation, particularly of the resource poor people, in the development process.

4. Smartphone/mobile: In the agricultural context, the cellular phones has empowered the farmers to communicate from local to administrative levels regarding the agricultural trade. With the help of smartphone it is easier to exchange the information about buying and selling inputs and farm commodities within the shorter period. This has reduced the cost of travelling and improved the production efficiency for farming communities living in far funk/remote areas. The mobile phone has tremendously been diffused even into the marginalized and underdeveloped farming communities because of its flexibility, affordability and user- friendly nature as compared to other ICT tools.

■■■■

Chapter - 10 Technology Demonstration

- Technology demonstration is the most effective way to show how a thing works, how to do the work, principles involved in an operation and to show the end results of the technology/methodology adopted.
- 'Believing through seeing' and 'learning by doing' accomplished through demonstrations help KVKs in technology integration. This process arouses interest and improves the adoption, because it is based on the principles of 'learning by doing' and 'seeing is believing'.
- Amongst the communication and educational techniques 'demonstration' are the oldest, best and simplest tools for transmitting sophisticated technology in a simple and understandable form.
- Demonstration means showing by doing. The basic principles of demonstration is learning by seeing and doing.

On the basis of purpose for which technology demonstration is conducted it is classified into:

1. Method demonstration

It is the short type demonstration. The method demonstration evolved by Dr. Seeman A. Knapp. This is a single practice demonstration *i.e.*, how to apply fertilizer in the field. This cannot compare between old and new technique or skill. It's main purpose to provide only skill. It is oldest form of teaching.

A method demonstration is conducted to explain how to carry out a particular operation according to its principles so that it is carried out systematically and yields better result. Through method demonstration we teach the learners about how to do something. The learners will be shown the right method for doing an old practice or taught about a new practice that are introduced to them. The method demonstration teaches new skill and helps the learner to obtain practical knowledge about something they need to practice or do in their day to day life.

Some of the specific purposes for which the demonstration is useful are urea molasses treatment of paddy straw, compost pit, biogas plant, pesticide spraying, milking operation, sanitation methods, seed treatment, milk collection and storage, de-beaking of poultry, egg preservation, cattle feed preparation, operation of improved implements, sericultural operation, beekeeping, grooming of cattle, dehorning, vaccination of poultry and cattle, semen collection, seed bed preparation, judging and selection of dairy and work cattle, hay making, silage making etc.

There are some points which should be taken into consideration while demonstrating practice.

Phases of demonstration

An effective demonstration programme involves 5 essential phases:

1. The **first phase** is of analyzing the situation.
2. The **second phase** is deciding on goals to be accomplished in conducting a demonstration.
3. The **third phase** is the demonstration plan and execution of demonstration.
   a. Pre-demonstration stage
   b. Execution stage
   c. Follow up or post demonstration stage
4. The **fourth phase** is evaluation of the demonstration programme.
5. The **fifth phase** is reconsidering after evaluation has taken place.

Important tips conducting a good demonstration

i. The demonstrator should encourage the audience and see that people understand and learn to do what is being demonstrated.

ii. Before starting the actual working, he should tell the audience.

iii. Tell the audience especially the interest of farmers.

iv. Help the audience to go through the steps of demonstration so that they can repeat the process themselves without further any guidance.

v. A demonstrator should be well prepared in the subject himself and should take the help of his colleagues and teachers, whenever necessary.

vi. Invite questions and create an atmosphere that the audience may clarify it doubts.

2. Result demonstration

The result demonstrations are shown in comparison to existing practices. It is long type of demonstration. In result demonstration, the extension worker may take the help of audio-visual aid like film, slides, pictures and comparative diagrams etc. So the farmers imagination is fired and they have some idea of how things were done to obtain higher yields. The value of a new practice can be realized better by seeing the end product or outcome in comparison with the existing practice.

In extension education the result demonstration serves as an important tool to convince the farmer about the value of a new idea or innovation that are introduced to them as an option to their existing practice. Unless the farmers see the outcome or results of the recommended practice in comparison with their existing practice with their own eyes and experience, it becomes difficult for the extension worker to make the farmer to adopt the recommended variety or practice. In such circumstances, the result demonstration helps the extension worker to make his job easy. The result demonstrations may be used for a single recommended practice or a series of practices that come in sequence with respect to a problem.

The framework, observability of results, trialability, complexity of operation and compatibility to the local people are the essential points to be considered for deciding a result demonstration. For example, it is easy to conduct the result demonstration for showing an increased yield of a hybrid seed in comparison with a local variety than to show the increased milk yielding potential of a crossbred cow with that of an indigenous one. The time framework required for the former operation is a maximum of three months whereas a minimum period of three years are necessary to show the difference in the milk production and other reproductive parameters. Some of the recommended jobs for which result demonstration is useful are improved production technologies, high yielding crop varieties, improved seed, fertilizer application, plant protection measures, irrigation measures, improved tools and instruments etc.

Important points for result demonstration

i. The students should demonstrate on specific problems of the line sowing of wheat against broadcast, hybrid maize against local variety.

ii. It should be carried out correctly, in detail and must be in sequence.

iii. Better to conduct more than one demonstration in the same village, on the same subject.

iv. Never try to discover new truths.

v. Do not repeat demonstration.

vi. Use audio-visual aids to support the result demonstration

vii. The demonstration should be located by a well-travelled roadside or path.

viii. It is very effective in adoption.

ix. It is used to show the superiority of practices.

x. It is constructed by a farmer under direct supervision of an extension worker.

From the above discussion it is understood that result demonstration is:

1. The method to show the worth or end product or outcome of a practice or an idea.
2. Conducted mainly to show the differences between two practices especially when one is considered more superior than the other in giving the result or outcome.
3. To compare the results. For comparisons, records are to be maintained.

3. National or composite demonstration

A composite demonstration is a combination of field based result demonstrations and a chain of skill oriented method demonstrations. For example, combined effect of irrigation and fertilizer application on grain yield and quality of a newly released crop varieties may be demonstrated as a composite demonstration.

i. It is a combined form of both method and result demonstration.

ii. It is first line demonstration or front line extension project.

iii. It is conducted by the researchers on the farmer's field.

iv. The main objective is to show how production can be increased per unit area and per unit time.

v. Multiple cropping system are usually followed and high yielding varieties are used in this demonstration.

vi. National demonstration was first time started in 1965 by Ministry of Food and Agriculture.

vii. This programme was given to ICAR in 1967.

viii. At present it is become a part of agricultural production programme.

Front-Line Demonstration

Front-Line Demonstration (FLD) is a concept of field demonstration evolved by the Indian Council of Agricultural Research during the inception of Technology Mission on Oilseed Crops during mid-eighties. The field demonstrations conducted under the close supervision of scientists of the National Agriculture Research System is called front-line demonstrations because the technologies are demonstrated for the first time by the scientists themselves before being fed into the main extension system of the State Department of Agriculture. KVKs conduct frontline demonstration only on those technologies suited to their micro location as proved by their own assessment and refinement through OFTs.

The main objective of Front-Line Demonstrations is to demonstrate latest crop production technologies and its management practices in the farmers' field under different agro-climatic regions and farming situations. While demonstrating the technologies in the farmers' field, the scientist are required to study the factors contributing towards higher crop production, field constraints of production and thereby generate production data and feedback information. Front-Line Demonstrations are conducted in a block of two or four hectares land in order to have better impact of the demonstrated technologies on the farmers and field level extension functionaries.

Comparison between FLD v/s Other Demonstrations

The Front-Line Demonstrations are different from the normal demonstrations conducted by the extension functionaries. The specific features of Front-Line Demonstrations are:

- Front-Line Demonstrations are conducted under the close supervision of the scientists of the National Agriculture Research System comprising of Krishi Vigyan Kendras, ICAR Institutes, National Research Centres, Project Directorates and the State Agricultural Universities and its Regional Research Stations.
- Only latest proven technologies are selected by KVKs for Front-Line Demonstrations.
- Front-Line Demonstrations are organized in a block of two to four hectares involving all those farmers whose plots fall in the identified demonstration block.
- Only critical inputs and training are provided from the scheme budget, remaining inputs are supplied by the farmers themselves.

- Training of the farmers associated with the Front-Line Demonstrations is a prerequisite for conducting such demonstrations.
- The target audience of the Front-Line Demonstration is both farmers and the extension officers. The purpose is to convince extension functionaries and farmers together about the potentialities of the technologies for further wide scale diffusion.
- Front-Line Demonstrations are used as a source of generating data on factors contributing higher crop yields and constraints of production under various farming situations.

Types of Frontline Demonstrations

Frontline demonstration is a long term educational activity conducted in a systematic manner in farmers' fields to show worth of a new practice/technology. "Seeing is believing" is the basic philosophy of frontline demonstrations. Only proven technologies (based on OFT results) are selected for field demonstrations. Field demonstrations educate farmers through results obtained in terms of varieties resistant to disease and pest, quality of the grains and overall higher yields. In addition, it also educates the farmers in term of input-output ratio and economic gains in terms of net returns. Basically the frontline demonstrations are different types:

I. Single practice FLD

A 'Single Practice FLD' aims at proving the worth of a single practice such as effect of balanced fertilizers in rice crop, higher yields from newly released varieties of hybrid maize, effect of irrigation at different stages of crop growth, effect of new pesticide against control of insects/pests etc.

II. Composite practice FLD

A 'Composite FLD' is a combination of field based result demonstrations and a chain of skill oriented method demonstrations. A long term sequential method demonstration (Composite FLD) aims at demonstrating the superiority of a package of practice in growing a field crop. The effect of one practice in harnessing the effect of other practices is also demonstrated and studied. For example, combined effect of irrigation and fertilizer application on grain yield and quality of a newly released varieties of crops may be demonstrated as a composite FLD. Many times, there is an over emphasis on FLD in KVK work. Successful FLDs lead to higher adoption of demonstrated practices by the farmers as they develop the confidence amongst them in the practices demonstrated.

Field demonstrations provide an effective learning situation as farmers as well as extension personnel of line departments "See the crops themselves", "interact with the KVK scientists on the fields", and "get doubts clarified on the spot". It is, therefore, essential that all FLDs should be well planned and executed leaving no room for failure. KVK Scientists; therefore, have to be very careful in planning and conducting the FLDs. Without specific purpose, demonstrations should not be organized. Demonstration should be conducted only when situation demands.

III. Cluster frontline demonstration

Cluster frontline demonstration to demonstrate the production potential of newly released technologies on the farmer's field at a different location in a given farming system and organized farming and extension activities for farmer and extension workers for the dissemination of various technologies.

Cluster frontline demonstration is a novel approach to provide a direct interface between researcher and farmer for the transfer of technologies developed by them and to get direct feedback from farming community.

■■■■

Chapter - 11 Adoption and Diffusion of Innovations

Adoption: Adoption is a decision to make full use of an innovations as the best course of action available.

Diffusion: Diffusion is the process by which an innovation is communicated through certain channels over time among the members of a social system.

Innovation: An innovation is an idea, practice or object that is perceived as new by an individual or other unit of adoption. The technologies, practices developed through research are innovations. These may be new crop varieties, new breed of livestock, new medicines and chemicals, new package and practices etc. Farmers themselves may develop some new practices which are also innovations.

Is an idea, practice or object that is perceived as new by an individual or group (or organisation).

Communication: The process by which participants create and share information to one another in order to reach a mutual understanding.

Time: Time involved in the innovation-decision process, the time taken to adopt an innovation by the adopter and the adoption rate across the social system.

Social system: Are a set of interrelated social units (e.g. individuals, informal groups, organisations) that are engaged in problem solving to achieve a common goal. It determines the boundary for a diffusion process; it can be affected by norms, and the degree to which individuals can influence one another

Diffusion of Innovations Theory

1. Rate of adoption is a key feature of the theory – Everett Rogers developed adopter categories to 'measure' innovativeness of farmers to produce a statistical model (normal distribution curve) to show the distribution of the five adopter categories over the average time of adoption (Fig. 1).

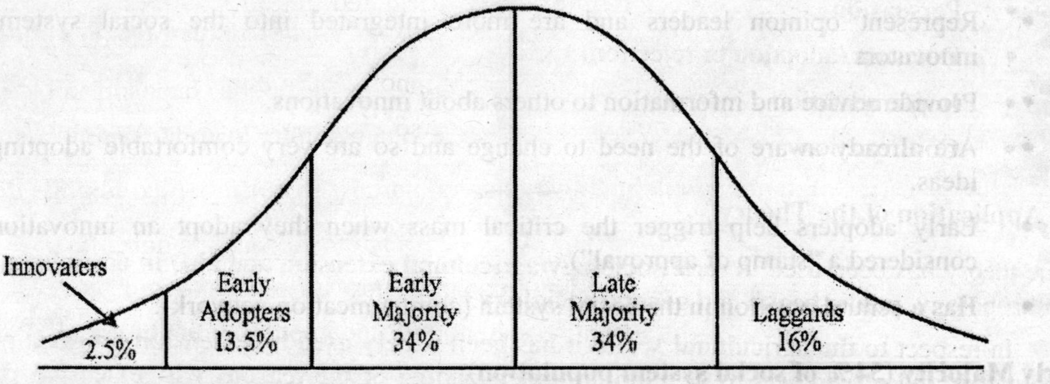

Fig. 1. Adopter categorizations on the basis of innovativeness (Rogers, E.M., 2003).

Everett M. Rogers is considered a founder of the Diffusion of Innovations theory. Rogers undertook a PhD (doctoral dissertation) in 1957 analysing the diffusion of several agricultural innovations in a rural community in Iowa. Rogers was convinced that the adoption of innovations follows a universal process of social change. It originated in communications to explain how, over time, an idea or product gains momentum and spreads (or diffuses) through a specific population or social system.

There are four main interacting elements : Diffusion of Innovations –

1. an innovation,
2. communicated through certain channels,
3. over time,
4. among members of a social system.

These four elements are present in every diffusion research study and in every diffusion program (*i.e.* the main elements are the variables in the diffusion process).

Rogers adoption categories explained

Innovators (2.5% of social system population)

- A "venturesome" approach to change.
- Quick to take up new ideas, knowledge and technologies.
- Cope with uncertainty and failure.
- Risk-takers.
- Play an important role in introducing innovations into the system and plays a gatekeeper role in the flow of information in a social system.

Early Adopters (13.5% of social system population)

- "Respected" members of the system.

- Represent opinion leaders and are more integrated into the social system than innovators.
- Provide advice and information to others about innovations.
- Are already aware of the need to change and so are very comfortable adopting new ideas.
- Early adopters help trigger the critical mass when they adopt an innovation (*i.e.* considered a "stamp of approval").
- Has a central position in the social system (communication network).

Early Majority (34% of social system population)

- "Deliberate" need to see evidence, think about and be convinced by others before being willing to adopt.
- Adopt innovations before the average person.
- Rarely perform leadership roles.
- Serve as important links in the diffusion process as they are the connection between very early and the relatively late adopters.

Late Majority (34% of social system population)

- "Sceptical" cautious about change and have a questioning attitude to innovations.
- Adopt new ideas just after the average person.
- Adoption may result from peer pressure or economic necessity rather than motivation for change.
- Innovation must be well supported by social norms to be desirable.
- Possibly have few resources therefore most of the uncertainty of the innovation must be removed before the late majority feel comfortable with adopting innovations.

Laggards (16% of social system population)

- "Traditional" they are bound by tradition and are very conservative.
- They are not opinion leaders and mostly considered 'isolates' in a social network sense (not connected strongly or to many other system members).
- The points of reference for laggards is the past.
- They are very sceptical of change and are the hardest group to motivate to adopt innovations.
- They take a lengthy amount of time to adopt in association with awareness of innovation.
- They demonstrate resistance towards innovations and are risk averse.

- They are in a vulnerable economic situation, therefore access to resources is constricted.
- Blame for not adopting can be located at the individual and system level.

Rogers also listed a series of correlations that show the positive relationship between a series of social variables and the likelihood of being an innovator type. Examples of these include:

- Education
- Literacy
- Higher social status
- Large-size farms
- Commercial economic orientation
- More favourable attitude to credit
- More favourable attitude to change
- More favourable attitude to education
- Intelligence
- Social participation
- Urban contacts
- Change agent contact
- Mass media exposure
- Exposure to interpersonal channels
- More active information seeking
- Knowledge of innovations, and
- Opinion leadership.

The implication of adopting the theory is that extension professionals/practitioners need to develop different practices/messages for different 'types' of adopters (*i.e.* farmer segmentation).

There are five main characteristics of innovations that determine how an innovation will be responded to by a potential farmer/end-user:

1. **Relative Advantage** – The degree to which an innovation is seen as better than the idea, program, or product it replaces.
2. **Compatibility** – How consistent the innovation is with the values, experiences, and needs of the potential adopters.
3. **Complexity** – How difficult the innovation is to understand and/or use.
4. **Trialability** – The extent to which the innovation can be tested or experimented with before a commitment to adopt is made.
5. **Observability** – The extent to which the innovation provides tangible results.

Rogers proposed a set of stages in decision-making in adopting an innovation:

- Knowledge

- Persuasion
- Decision (adoption or rejection)
- Implementation
- Confirmation

2. Application of the Theory

The theory has been used in rural sociology/agricultural extension and also in disciplines such as anthropology, public health and general sociology.

In respect to the agricultural sector it has been widely used by extension program planners, evaluators and researchers to develop an understanding of the reasons why extension programs result in adoption or rejection of a particular new practice. It also provides a general understanding of the impact of extension programs through the extent/degree of innovation adoption.

Many governments of developing economies have used the Diffusion of Innovations theory to shape the conceptual framework and implementation design of international rural development programs.

3. Critique of the Theory

Publicised criticisms of the theory started to appear in the 1970s in the context of international development projects (Rogers, 2003). The key criticism was that innovations were being targeted to the "Innovators" and "Early Adopters" – the more 'progressive' farmers, with the expectation that innovative practices would trickle down to the majority of farmers. However, the reality was that the application of the theory was viewed as a source of inequity, dividing rural communities and not benefiting/assisting those in most need. This was particularly noticeable when the diffusion of innovations process benefited larger farmers by increasing their production but decreasing the market prices/farm gate returns received by all farmers in the region (including the non-adopters).

4. Other limitations of the theory

- **A Pro-Innovation Bias** – There is the implication that an innovation should be diffused and adopted by all farmers. The act of innovating is considered positive and the act of rejecting an innovation is considered negative.
- **Individual-Blame Bias** – The development agency is not blamed for its lack of response to the needs of farmers. Rather, the individuals who do not adopt the innovation are blamed for their lack of response.
- **Issue of Equality** – The negative impacts of the theory are not considered. What are the consequences in terms of unemployment, migration of rural people and equitable distribution of incomes? Will the innovation widen or narrow socioeconomic gaps?
- **Bias in Favour of Larger and Wealthier Farmers** – Innovation investors tend to target the already innovative, successful and information-seeking farmers who have the capacity to resource themselves to experiment and adopt an innovation – which can

lead to great productivity/profitability, consequently the social divide amongst farmers widens as a result.

Limitations of diffusion research

- It tends to assume innovations originate at research institutes/central agency rather than farmers themselves, other players in the RD&E system.
- It assumes adoption of innovation is always desirable (normative), it tends not to evaluate innovations from an end-user perspective.
- There is a risk of technical/research messages being lost in translation to the farmer because of the language used.
- It assumes there is enough research information available to the extension/change agent and does not tend to see knowledge as a combination of research outputs with farmer's knowledge, experience and interpretation of the problem.
- There are few systematic evaluations of the adoption and diffusion model, research on model does not tend to focus on systemic change (changes to the social system), rather the focus is on discrete technical changes, changes by individuals and groups rather than institutions and societies as a political.

Diffusion of innovations

Diffusion is the process by which an innovation is communicated through certain channels over time among the members of a social system. It is a special type of communication, in that the messages are concerned with new ideas.

Elements in the diffusion of innovations

The four main elements in diffusion of innovations are:

1. The innovation

An innovation is an idea, practice or object that is perceived as new by an individual or other unit of adoption. The perceived newness of the idea for the individual determines his or her reaction to it. The "newness" aspect of an innovation may be expressed in terms of knowledge, persuasion or a decision to adopt.

In this context, to know about the **perceived attributes of innovation** would be appropriate which are described below:

a. **Relative advantage:** It is the degree to which an innovation is perceived as better than the idea it supersedes. The degree of relative advantage may be measured in economic terms, but social-prestige factors, convenience and satisfaction are also often the important components.

b. **Compatibility:** It is the degree to which an innovation is perceived as being consistent with the existing values, past experiences and needs of potential adopters.

c. **Complexity:** It is the degree to which an innovation is perceived as difficult to understand and use. In general, new ideas that are simpler to understand will be adopted more rapidly than innovations that require the adopter to develop new skills and understandings.

d. **Trialability:** It is the degree to which an innovation may be experimented with on a limited basis. An innovation that is trialable represents less uncertainty to the individual who is considering it for adoption, as it is possible to learn by doing.

e. **Observability:** It is the degree to which the results of an innovation are visible to others. The easier it is for individuals to see the results of an innovation, the more likely they are to adopt.

2. Communication channels

A communication channel is the means by which messages get from one individual to another. The following classification of channels would help the communicator to use them appropriately:

i. **Interpersonal channels** - It refers to those which are used for face to face communication between two or more individuals.

ii. **Mass media channels** - These enable the messages to reach a larger, diverse audience simultaneously in a relatively shorter time. e.g.: Radio and T.V.

iii. **Localite channels** - They originate within the social system of the receiver. eg: neighbours, relatives, opinion leaders etc.

iv. **Cosmopolite channels** - They originate outside a particular social system. eg: extension worker, sales personnel etc.

3. Time

It is an important element in the diffusion process. Time is an obvious aspect of any communication process. Time does not exist independently of events, but it is an aspect of every activity. The time dimension is involved in diffusion (i) in the innovation - decision process, (ii) in the innovativeness of an individual or other unit of adoption, and (iii) innovation's rate of adoption in a system.

4. Social System

It is defined as a set of interrelated units that are engaged in joint problem solving to accomplish a common goal. The members or units of a social system may be individuals, informal groups, organisations and/or subsystems. The social system constitutes a boundary within which an innovation diffuses.

INNOVATION - DECISION PROCESS

As an alternative to the "Stages in the adoption process" *viz.*, Awareness, Interest, Evaluation, Trial and Adoption, due to the advancements in diffusion research, currently" Innovation - Decision process" is proposed which enlightens the sequential stages in the adoption - decisions made by individuals or other units of adoption.

The "Innovation - Decision Process" is the process through which an individual (or other decision - making unit) passes from first knowledge of an innovation, to forming an attitude towards the innovation to a decision to adopt or reject, to implementation of the new idea, and to confirmation of this decision.

This process consists of a series of actions and choices over time through which an individual or an organisation evaluates a new idea and decides whether or not to incorporate the new idea into on going practice.

The conceptualization of the model of the innovation decision process consist of the following five stages. These are:

1. Knowledge stage

Knowledge occurs when an individual (or the decision - making unit) is exposed to the innovation's existence and gains some understanding of how it functions.

The following three types of knowledge possessed by an individual influence the decisions :

i. **Awareness - knowledge** - motivates an individual to seek "how-to" knowledge and principles knowledge. This type of information - seeking is concentrated as the knowledge stage of the innovation - decision process, but it may also occur at the persuasion and decision stages.

ii. **How-to knowledge** - consists of information necessary to use an innovation properly. When an adequate level of how-to knowledge is not obtained prior to the trial and adoption of an innovation, rejection or discontinuance is likely to result. Change agents could perhaps play their distinctive role to concentrate on "how-to knowledge" at the trial and decision stage in the process.

iii. **Principles knowledge** - consists of information dealing with the functioning principles underlying how the innovation works. It is usually possible to adopt an innovation without principles knowledge, but the danger of misusing the new idea is greater, and discontinuance may result. The long-range competence of individuals to judge future innovations is facilitated by principles knowledge.

2. Persuasion stage

Persuasion occurs when an individual (or other decision - making unit) forms a favourable or unfavourable attitude toward the innovation.

While the mental activity as the knowledge stage was mainly cognitive (or knowing), the main type of thinking at the persuasion function is affective (or feeling). At this stage, a general perception of the innovation is developed. The individual becomes more psychologically involved with the innovation and hence he or she seeks information about the new idea.

3. Decision stage

Decision occurs when an individual (or other decision - making unit) engages in activities that lead to a choice to adopt or reject the innovation.

Adoption is a decision to make full use of an innovation as the best course of action available. Rejection is a decision not to adopt an innovation.

The small - scale trial is often part of the decision to adopt, and is important as a means to decrease the perceived uncertainty of the innovation for the adopter.

4. Implementation stage

Implementation occurs when an individual (or other decision - making unit) puts an innovation into use. Until the implementation stage, the innovation-decision process has been a strictly mental exercise. But implementation involves overt behaviour change as the new idea is actually part into practice.

Problems of implementation are likely to be more serious when the adopter is an organisation rather than an individual. Reason is that in an organisational setting, a number of individuals are usually involved in the innovation - decision process, and the implementers are often a different set of people from the decision makers.

5. Confirmation stage

Confirmation occurs when an individual (or other decision - making unit) seeks reinforcement of an innovation - decision already made, but he or she may reverse this previous decision if exposed to conflicting messages about the innovation.

The confirmation stage continues after the decision to adopt or reject for an indefinite period in time. At this stage, the change agents have the additional responsibility of supporting messages to individuals who have previously adopted.

As a sequential effect, there is a possibility for "discontinuance". A discontinuance is a decision to reject an innovation after having previously adopted it. There are two types of discontinuances:

i. **Replacement discontinuance -** is a decision to reject an idea in order to adopt a better idea that supersedes it.

ii. **Disenchantment discontinuance -** is a decision to reject an idea as a result of dissatisfaction with its performance.

ADOPTER CATEGORIES

There are different categories of farmers. According to Rogers (1971), the farmers based on their innovativeness can be classified as

1. Innovators (Venturesome)
2. Early adopters (Respectable)
3. Early majority (Deliberate)
4. Late majority (Skeptical)
5. Laggards (Traditional)

1. Innovators (Venturesome)

Observers have noted that venturesomeness is almost an obsession with innovators. They are eager to try new ideas. This interest leads them out of a local circle of peers and into more cosmopolite social relationships. Communication patterns and friendships among a clique of innovators are common, even though the geographical distance between the innovators may be great. Being an innovator has several prerequisites. These include control of substantial financial resources to absorb the understand and apply complex technical knowledge.

The salient value of the innovator is venturesomeness. He desires the hazardous, the rash, the daring, and the risky. The innovator also must be willing to accept an occasional setback when one of the new ideas he adopts proves unsuccessful.

These are the first people to adopt a new idea, much ahead of other people. They are very few in numbers, probably not more than one or two in a community.

Specific characteristics of innovator

1. Have larger farms.
2. High net worth and risk capital.
3. Willing to take risks.
4. Usually not past middle age
5. Generally well educated
6. Have respect and prestige in progressive communities but not in conservative type of communities.
7. Mentally alert and actively seeking new ideas.
8. Their sphere of influence and activity often goes beyond the community boundaries.
9. They have many formal and informal contact outside the immediate locality.
10. They often by-pass the local extension worker in getting information from the originating sources, and may learn about new things even before he does. They sometimes manage to get samples of seeds or chemicals even before they are released for public use.
11. They subscribe to many farm magazines and specialised publications.
12. Other farmers may watch the innovators and know what they are doing but the innovators are not generally named by other farmers as "neighbours and friends" to whom they go for information.

2. Early Adopter (Respectable)

Early adopters are a more integrated part of the local social system than are innovators. Whereas innovators are cosmopolites, early adopters are localities. This adopter's category, more than any other, has the greatest degree of opinion leadership in most social systems. Potential adopters look to early adopters for advice and information about the innovation. The early adopter is considered by many as "the man to check with" before using a new idea. This adopter category is generally sought by change agents to be a local missionary for speeding the diffusion process. Because early adopters are not too far ahead of the average individual in innovativeness, they serve as a role model for many other members of a social system. Members of a social system respect the early adopter. The early adopter is respected by his peers. He is the embodiment of successful and discrete use of new ideas. Early adopter knows that he must continue to earn this esteem of his colleagues if his position in the social structure is to be maintained.

Specific characteristics of early adopters

1. Younger than those who have a slower adoption rate, but not necessarily younger than the innovators.
2. They are not the persons who test the untried ideas but they are quickest to use tried ideas in their own situations.
3. Have large farms.
4. Higher education than those who adopt more slowly.
5. High income.
6. They participate more in the format activities of the community.
7. They also participate more in government programmes.
8. This group usually furnishes a disproportionate amount of the formal leadership (elected positions) in the community.
9. They read papers and farm journals and receive more bulletins than people who adopt later.
10. They may be regarded as community adoption leaders.

3. Early Majority(Deliberate Local Adoption Leaders)

The early majority adopt new ideas just before the average member of a social system. The early majority interact frequently with their peers, but leadership position; are rarely held by them. The early majority's unique position; between the very early and relatively late to adopt make; them an important link in the diffusion process.

The early majority may deliberate for some time before completely adopting a new idea. Their innovation-decision is relatively longer than that of the innovator and the early adopter. "Be not the last to lay the old aside, nor the first by which the new is tried", might be the motto of the early majority. They follow with deliberate willingness in adopting innovations, but seldom lead.

Specific characteristics of early majority

1. Slightly above average in age, education and farming experience.
2. They take a few more farm journals and bulletins than the average.
3. They have medium high social and economic status.
4. Less active in formal groups than early adopters, but more active than those adopting later.

5. In many cases, they are not formal leaders in the association
6. They also attend extension meetings and farm demonstrations.
7. They are most likely to be informal resources than early adopters and innovators, and so cannot afford to make hasty or poor decisions.
8. They associate mainly with people of their own community.
9. They are highly the opinions their neighbours and friends hold about them; for this is their main source of status and prestige.
10. They are mostly mentioned as "neighbours and friends" from whom the majority of farmers seek information.

4. Late Majority (Skeptical)

The late majority adopt new ideas just after the average member of a social system. Adoption may be both an economic necessity and the answer to increasing social pressures. Innovations are approached with a skeptical and cautions air, and the late majority do not adopt until most other in their social system have done so. The weight of system norms must definitely favour the innovation before the late majority are convinced. They can be persuaded of the utility of new ideas, but the pressure of peers is necessary to motivate adoption.

Specific characteristics of late majority

1. Those in this group have less education and are older than the early majority.
2. They form the major part (formal organisational membership, although they participate less in such formal groups.
3. They take fewer leadership roles than the earlier adopters.
4. They take and read fewer papers, magazines and bulletins, than the early majority.
5. They do not participate in as many activities outside the community as do people that adopt earlier.

5. Laggards (Traditional)

Laggards are the last to adopt an innovation. They possess almost no opinion leadership. They are the most localite in their outlook of all adopter categories, many are near isolates. The point of reference for the laggard is the past. Decisions are usually made in terms of what has been done in previous generations. This individual interacts primarily with others who have traditional values. When laggards finally adopt an innovation, it may already have been superseded by another more recent idea which the innovators are already using. Laggards tend to be frankly suspicious of innovations, innovators, and change agents. Their tradition direction slows the innovation decision process to a crawl. Adoption lags far behind knowledge of the idea. Alienation from a too-fast-moving world is apparent in much of the laggard's outlook. While most individuals in a social system are looking to the road of change ahead, the laggards has his attention fixed on the rear-view mirror.

Specific characteristics of laggards

1. Least education.
2. Oldest.
3. Participate least in formal organisations, cooperatives and government programmes.
4. They hardly read farm magazines and bulletins.

Research attributes of innovation

The perceived attributes of innovations which are basic to extension are as follows:

1. **Relative advantage**: The degree to which an innovation is perceived as better than the idea it supersedes. The relative advantage may have a number of dimensions. The innovations which have more relative advantage are likely to be adopted quickly.
2. **Compatibility:** The degree to which an innovation is perceived as being consistent with the existing values, past experiences and needs of the potential adopters.

   Compatibility has two dimensions:

   (a) **Situational compatibility:** When a new crop variety suits the agro-climatic condition of the farmers, it indicates situational compatibility.

   (b) **Cultural compatibility:** When a breed of livestock advocated to the farmers is in agreement with their beliefs and values, it is cultural compatibility.
3. **Complexity:** The degree to which an innovation is perceived as difficult to understand and use. An innovation should as far as possible, be less complex for the farmer to understand and use. Complex technologies often require complementary adoption.
4. **Triability:** The degree to which an innovation may be experimented on a limited basis. Earlier adopters appears to be more concerned about the triability of an innovation than later adopters.
5. **Observability:** The degree to which the results of an innovation are visible. The visible impact of an innovation facilitates its diffusion in the social system.
6. **Predictability:** Predictability refers to the degree of certainty of receiving expected benefits from the adoption of an innovation.

The adoption process

There are five stages of adoption process.

1. Awareness; 2. Interest; 3. Evaluation; 4. Trial; 5. Adoption

1. **Awareness stage:** The individual learns of the existence of the new idea but lacks information about it.
2. **Interest stage:** The individual develops interest in the innovation and seeks additional information about it.
3. **Evaluation stage:** The individual makes mental application of the new idea to the present and anticipated future situations and decides whether or not to try it.

4. **Trial stage:** The individual actually applies the new idea on a small scale in order to determine its utility in own situation.
5. **Adoption stage:** The individual uses the new idea continuously on a full scale.

According to some scientists, the stages of adoption are dynamic and not static. The same five stages do not occur with all the adopters and all the practices. Sequences is not always the same. Some times one stage appears more than once. In some cases some stages are so short as to be imperceptible, and in other cases some stages seem to be skipped. The scheme of stages according to them are:

1. Need; 2. Awareness; 3. Interest; 4. Deliberation; 5. Trial; 6. Evaluation; 7. Adoption.

■■■■

Chapter - 12 Extension Programme Planning

Extension programme planning is a decision making process.

Proper planning of the program is very important for achieving the success in extension work.

Programme planning is a procedure of working with the people in an effort to recognize unsatisfactory situations and to determine possible solutions or objectives and goals.

"An extension programme is a statement of situation, objectives, problems and solutions"; An extension programme is the sum total of all the activities and in undertakings of a country extension service. It includes:

1. The programme planning process
2. Written programme statements
3. Plan of work
4. Programme execution
5. Results
6. Evaluation

Program planning is a process through which representatives of the people are intensively involved with extension personnel and other professional people in four activities (Boyle, 1965).

a. Studying facts and trends.
b. Identifying problems and opportunities based on these facts and trends.
c. Making decision about problems and opportunities that should be given priority.
d. Establishing objectives or recommendation for future economic and social development of a community through educational programs.

Extension programme planning is the process of determining, developing and executing programs. It is continuous process whereby farm people with the guidance and leadership of extension personnel attempt to determine analysis and solve local problems. In this there are three characteristics:

a. What needs to be done.
b. When it should be done.
c. How it should be done (Musgraw, 1962).

Component of programme planning

The component of programme planning are complex. These are:

- People their needs.
- Useful technology.
- The educational process.
- Analysing situation.
- Making decisions about what should be done.
- Determining action.
- Projection the desired shape of things into the future etc.

Objectives of programme planning

The objectives of an extension programme is to influence people to make changes in their way of life and in making a living better.

The following objectives of having a programme may be stated as:

1. To ensure careful consideration of what is to be done and why.
2. To ensure continuity during changes in personnel.
3. To furnish a guide against which to judge all new proposals.
4. To help develop leadership.
5. To develop a common understanding about the means and ends behaviour various functionaries and organizations.
6. To have available in written form a statement for public use.
7. To establish objectives towards which progress can be measured and evaluated.
8. To avoid waste of time and money and to promote efficiency.
9. To have a means of choosing the important from incidental problems and the permanent from the temporary changes.
10. To justify expenditure and to ensure flow of funds.

Principles of extension programme planning

For the development of a useful and effective programme, the following principles should be kept in mind, which is suggested by Kelsey and Hearne (1955).

Principles related to extension program

1. Program planning is based on analysis of the facts in the situation.
2. Program planning selects problems based on needs and interest of local people.
3. Program planning determines definite objective and solutions, which offer satisfaction.
4. Program planning has permanence width flexibility.
5. Program planning has balance with emphasis.

Principles related to planning process

6. Program planning has a definite plan of work.
7. Program planning is an educational process.
8. Program planning is a continuous process.
9. Program planning is a coordinating process.
10. Program planning involves local people and their institutions.
11. Program planning provides for evaluation of results.

A principle is a statement of policy to guide decision and action in a consistent manner (Mathews). Principle is a universal truth that has been observed and found to be true under varying conditions and circumstances. A principle is a fundamental truth and a settled rule of action.

Extension programme have the definite purpose of improving rural life through individual group and community action. Extension programme planning has certain principles which hold irrespective of the nature cliental and the enterprises they may pursuing.

1. Extension programme planning should be an analysis of the past experiences, present situation and future needs: For programme determination, adequate information about the people and their situation has to be collected. The present situation is to be analysed and interpreted on the basis of past experiences by taking local people into confidence. This shall help in arriving at the future needs.

2. Extension programme should have clear and significant objectives which could satisfy important situation and future needs: The ultimate objectives of program building are to satisfy the needs of the people. For this purpose, significant objectives pertaining to important needs of the people should be selected and clearly stated. The emphasis shall be on what is attainable rather than on what is ideal; also one should not lose sight of the later.

3. Extension programme should be fixed up priority on the basis of available resources and time: The rural people particularly in the developing countries have a multitude of problem. All problems cannot be taken up at a time for solution because of the limitation of trained personnel, availability of funds, facilities and other resources. Time is also a limiting factor as both the people and the funding agencies cannot wait for an indefinite period of time to get the result. Considering all these parameters, it is essential to fix up priorities in the programme.

4. Extension programs should clearly indicate the availability and utilization of resources : An extension programme should clearly state where from the funds, facilities, supplies and the needed personnel shall be available and how these shall be utilized. This shall make the program practical and workable.

5. Extension programs should have a general agreement at various level : Programs prepared at the various levels such as village, district, state and national levels should conform to each other and shall not work at cross purposes. Similarly the extension programme of a particular department should not be conflict or contradiction with the extension programme of another department.

6. Extension programme should involve people at the local level: Extension programs are implemented at the local level. Local people should therefore be involved all through from program formulation to programme implemented.

7. Extension programs should involve relevant institutions and organization: Extension programs cannot be implemented in isolation. It requires the support of many institution and organization. The program should broadly indicate the institutions and organizations to be involved and how they shall contribute in attaining the program objectives.

8. Extension programs should have definite plan of work: The plan of work may be separately drawn up or incorporated in the program. The program should at least broadly indicate how it will be executed. Unless the plan of work is drawn up the program remains a theoretical exercise.

9. Extension programs should provide for evaluation of results and reconsideration of the program: Extension program is not a static outline of activities. The program should make provision for periodical monitoring and evaluation of result to judge its progress. On the basis of the findings of evaluation, the program should be suitably modified to facilitate its reaching the objectives within the stipulated period of time.

10. Extension programs should provide equal distribution of benefits amongst the members of the community: It has been found that in a community, generally the resources rich person benefit more in comparison to the resources poor from the implementation of extension programs. As this may generate social disparity and social tensions. The planning of extension programs should give adequate emphasis on the weaker section of the community.

Importance of extension programme planning

The importance of extension programme planning are given below:

1. Programme planning makes complex work to simple.
2. Planning makes the programme decent and effective.
3. Development of village and people needs are dependent on effective planning.
4. It helps to follow the right course of action.
5. It is also effective in solving problem.
6. It helps in growth and expansion of business.
7. It ensures careful consideration of what is to be done and why.
8. It establishes objectives towards which progress can be measured and evaluated.
9. It develops common understanding about the means and ends between various functionaries and organizations.
10. It helps to develop leadership.
11. It avoids waste of time and money and promote efficiency.
12. It justifies expenditure and ensure flow of funds.
13. It helps to have available in written form a statement for public use.
14. To give continuity during changes of personnel.
15. To coordinates the efforts of the different people working for rural development.

16. To have a means of choosing the important from the incidental problems and the permanent from the temporary changes.

Characteristics of good program

A good program should have the following characteristics

1. The written program should be suitable for use by the staff, planning groups and other individuals or groups concerned with the program.
2. It should state the primary facts that clearly reveals with the situation on major subjects or problem areas.
3. It should clearly state that the important problems or needs identified by the staff and the people in the programming process.
4. It should clearly state both long term and short term objectives for each major subject and problem that is to be focused on in program execution over a period of time.
5. It should state the objectives of the program clearly and meaningfully.
6. It should specify the subject matter related to each objectives that is highly significant to people socially and economically.
7. It should include the summary of the long term program prepared in a form suitable for public distribution.
8. It should be made available in a summary form to all members of the planning groups and professional staff.
9. It should be circulated by appropriate means so that the general public can understand its nature and objectives.
10. It should be used as a basis for developing an annual plan of work.

Nature of extension program planning

1. Extension program planning is a process: The concept of process involves a method *i.e.* a process should be viewed as a sequential set of steps or several systematically ordered steps of planning, the performance of which leads to the accomplishment of a goal. In extension program planning, the immediate goal would be the development of a program document. Many programs program planning process takes place at any particular time and at different levels of extension organizations. For example, program planning occurs at a national level (5 years planning), at the state level state plans, annual plans and at block level.

2. Extension program planning is a decision making process: The scientific are put to value judgements of the people through implementation of a national planning model in order to decide a program which will be carried out through the extension teaching activities.

3. Extension program planning requires advance thinking: Good extension program planning is an intellectual activity since it usually involves a study of use facts and principles. It requires knowledge, imagination and reasoning ability. It is a complex exercise and involves people and their needs, their interest, useful technology, analyzing a situation and making decisions about what should be done, determining useful actions, projecting the desired shape of things in future and several other components which are rarely simple.

4. Extension program planning requires skills and ability on the part of planner: Effective program planning requires higher level professional skills, ability and understanding in the following broad areas:

i. Understanding the nature and role of extension education organization.

ii. Knowledge and understanding of the technology related to the subject for which the program is concerned.

iii. Ability to clarify, the objectives of a program and to state that they are useful in guiding its execution.

iv. Skills at seeing the relationship between principles and practices.

v. Skills in inquiry and human relationship.

5. Extension program planning is built around content: It is built around available improved technology, the people, their resources, problems, needs and interests. Without some expressed purpose, there can be no planning.

6. Extension program planning is a social action process: Extension program planning involves interaction and the decisions taken in the form of a program, which affect others. Extension planning involves specialists and people representatives in this process and the scientific data is put to value judgment so as to decide the intensive direction of change, appropriate methods to be used to reach the goals.

7. Extension program planning is a collaborative effort: Extension program planning is a collaborative effort involving identification, assessment or evaluation of needs, problems, resources priorities and solutions.

8. Extension program planning is a system: It is a system as it's procedure and process are interrelated, ordered and linked progressively to form a collective whole. It includes several processes such as planning, designing, implementing and evaluation etc.

9. The end product of extension program planning is an extension program: The first consideration for anyone who is to concern himself with a process or set of procedures for planning is to clearly identify the primary purpose of planning process to be developed. According to Vande Berg (1965), the primary purpose of any planning first and foremost is that developing a sound, defensible and progressive codes of action or 10 plans. In the process followed, many other benefits might occur, such as the education of participants but we want a plan, which can and will be used.

Types of program planning

A. According to duration

- Short term plan
- Medium term plan
- Long tern plan

B. According to extent of participation

- No participation
- Restricted participation

– Liberal participation

C. According to leadership approach

– Democratic approach

– Autocratic approach

D. According to the place of planning

– Centralised planning

– Decentralised planning

– Grass-root planning

E. According to development approach

– Isolated planning

– Integrated planning (Crop, livestock, fisheries, food and nutrition)

– Comprehensive planning (most advances planning or total planning)

Steps of extension programme planning

The principles of extension programme planning indicate earlier shall help us to develop the logical steps in program planning process. There are 8 steps which form a continuous cycle of program determination and programme implementation in extension. The steps are:

1. Collection of facts

It is the starting point of program planning process. Pertinent data may be collected from the available records and by survey of the area. Information related to the people their enterprises, levels of technology, facilities and constraints values etc. relevant to program building may be collected. Information also be collected from other organizations in the area. The program planning process is explain with an example. There is a village where farmers are poor and something is to be done to improve the condition. We conduct a survey of the village and collect information on the number of farm facilities, their occupations, land use pattern, utilization of water resources, facilities for marketing availability of inputs and credit, their attitude towards various enterprise, the strength of extension services in the area etc. We also collect relevant information from other organizations.

2. Analysis of situation

The data and information collected are then analysed in unbiased (few) way, keeping in view the feeling expressed by the client system. This shall help in understanding the situation in its proper perspective.

3. Identification of problems

A correct analysis and interpretation of the data shall help in correctly identifying problems. There may be many problems but only the urgent and significant ones which may be solved with the available resources and within limits of time should be selected. Selection of large number of problems which cannot be properly managed may lead to a failure of the program and generate frustration among the people. In the present example, after analysis of the data and relevant information with the local people three most important problems are identified which need immediate solution. For example, low levels of income employment and nutrition in the farm families.

4. Determination of objectives and goals

The objectives are then set forth on the basis of the significant needs identified. The objectives should be direct and stated in the clear terms. In the present examples, the objectives then become to increase levels of income, employment and nutrition of the farm families in the village.

To make the objectives realistic and actionable, there is need to state them in terms of specific goals. In the determination of goals it may be necessary to again go through the data and information analysed to findout what could actually be done in the existing situation with the available resources and time which will be compatible and with which the people cooperate. It is necessary to discuss with the local people and institution which shall also legitimize the program planning process.

5. Developing plan of work and calendar of operation

The plan of work should be in written form and shall indicate who shall do what job *i.e.* what the change agent system and thus line system shall do which institutions, organization, service department shall be involved what will be the financial requirement and how it shall be met; what arrangements shall be made for marketing of the produce, training of the farmers and so on. The plan should have all the essential details and no important should be left out.

The calendar of operation shall be prepared on the basis of the plan of work and it should specified when a particular work shall be done preferably mentioning data and time; how much quantity of different inputs including credit shall be required and when these must be made available when where and for how many days the farmers and farm women shall be trained. Who are the specialist to be involved in training and repairing the handouts when the publication shall be ready for distribution etc. that is the calendar of operation shall specifically state how and when all the significance activities shall be performed. This should be at least for one season or for a period of one year. In that case, they may be termed as seasonal plan or annual plan.

6. Follow trough plan of work and calendar

This is not a routine of work as many people may think. Training of participants, communication of information, conducting method demonstration, making regular visits and mentioning are some of the important function the extension agent shall perform at this stage. The work shall include solving and unforeseen problems and training cooperative steps where needed. The performance of extension agent and the organizational support received at this stage may make the difference between success or failure of a program. Obtaining feedback information as to what is happening to the farmers after introduction of new technology is extremely important at this stage.

7. Evaluation of Progress

Evaluation is the process of determining the extent to which we have been able to attain the objectives. All programs must have an inbuilt system of evaluation to know how well the work is done. It should be a continuous process not only to the measure the end result but also to ensure what all the steps are correctly followed. Evaluation may be formal and informal depending on the importance of the program and also on the availability to trained man power, funds, facilities and time. Program evaluation involves the following three essential steps.

- Setting up of some standards or criteria in relation to the objectives.
- Collection of information.
- Making judgement and drawing some unbiased and valid conclusion.

8. Reconsideration and revision of the program

On the basis of result of evaluation, the program should be reconsideration and revise if needed. This reconsideration should be done not only with the participants but also with the scientist, administrators in extension organization and local bodies. Reconsideration shall help in making necessary corrections and modifications in the program.

For the teaching plan the specific points should be included:

1. **Educational objectives:** Who is to be reached? *i.e.* audience. What subject matter is to be used? *i.e.* subject matter. What behavioural change is expected? *i.e.* behavioural change in the knowledge, skills, attitude and goals.
2. **Educational task:** For the subject matter content of each objective, here are specific task that must be identified in the plan.
3. Clientele to be reached must be identified the plan.
4. Methods to be used.
5. Location of teaching.
6. Calendar of teaching objectives.
7. The teacher.
8. Evaluation.

Difference between plan of work and calendar of work

	Plan of work		Calendar of work
1	Plan of work refers to abroad outline of activities to be done at each step of the program planning, execution and evaluation phases.	1	Calendar of work is the fragmentation of the annual plan of work.
2	It is not the part of the calendar of work.	2	It is the part of plan of work.
3	Usually a plan of work prepared for a year.	3	Usually a calendar of work is prepared for three months.

■■■■

Chapter - 13 Information and Communication Technology

Information and Communication Technology (ICT) can simply be defined in its simplest form as an electronic medium for creating, storing, manipulating receiving and sending information from one place to another. It makes message delivery faster, more convenient, easy to access, understand and interpret. It uses gadgets such as cell phones, the Internet, wireless network, computer, radio, television, Satellites, base stations etc. These resources are used to create, store, communicate, transmit and manage information.

Information and Communication Technologies (ICTs) is a broader term for Information Technology (IT), which refers to all communication technologies, including the internet, wireless networks, cell phones, computers, software, middleware, video-conferencing, social networking, and other media applications and services enabling users to access, retrieve, store, transmit, and manipulate information in a digital form.

ICTs are also used to refer to the convergence of media technology such as audio-visual and telephone networks with computer networks, by means of a unified system of cabling (including signal distribution and management) or link system. However, there is no universally accepted definition of ICTs considering that the concepts, methods and tools involved in ICTs are steadily evolving on an almost daily basis.

To define professional skill levels for its ICT professional education products, the IEEE Computer Society has adopted, for example, the Skills Framework for the Information Age (SFIA). The value of ICT strategies as a means of bridging the digital divide and as a powerful tool for economic and social development around the world should not be underestimated in agricultural and related sectors. Improving extension of ICT services to farmers would effectively improve the transmission of global open data for agriculture and nutrition for development of sensible solutions addressing food security, nutrition and sustainable agriculture issues.

By using ICTs, there have already been diverse types of innovations taking place in the agriculture sector, which include commodity and stock market price information and analysis, meteorological data collection, advisory services to farmers for agricultural extension, early warning systems for disaster prevention and control, financial services, traceability of agricultural products, agricultural statistical data gathering, etc.

Scope of Information and Communication Technology

- Information and Communication Technology is a field that has a wide coverage.

- It extensively deals with communication technology and how it impacts on other fields of human endeavor.
- It is the fastest growing academic field of study and a viable source of livelihood.
- It is the convergence of telephone and computer networking through a single cabling system with ease of data storage, manipulation, management and retrieval.
- It is concerned with database management, computer programming and software development.
- Web designing, mobile application development, project management, security, networking analysis, media equipment, computer engineering, computer studies, the internet, intranet, internet protocol (IP), system software, application software, signal technology, base station management etc.

Information and Communication Technology as a Field of Study

Information technology inclusion in most high schools curriculum is relatively new. However, it has gain prominence as some institutions have made it a compulsory subject. This is as a result of the understanding that it cut across every facet of human endeavour most especially the education sector. It is the fastest growing industry in the 21st century.

It is common place to hear e-teaching, e-learning, e-commerce, e-banking etc. It is, therefore incumbent on education curriculum developers to place ICT as a hub around which other disciplines revolve at least for the fact that it is a platform on which modern learning takes place.

Information and Communication Technology (ICT) has emerged as an effective facilitator in the development of any society and is a prime driving force in the growth of economies world wide. In this context, ICT have lot to offer to develop rural sector.

The most conspicuous development in the globalization context has been the information and communication technology (ICT) influx. It seems to have turned the world into a global village. Hence, the developing world is gradually catching up with the technological advancement to solve its socio-economic problems.

Methodology

a. **Key concepts**

1. **ICT (Information and Communication Technology):** "ICT as the set of activities that facilitates by electronic means, the capturing storage processing, transmission and display of information".
2. **Rural Development:** Rural development means, to promote the participation of rural people in ICT projects and programmes designed to assist their development and develop their information. Education and communication base of rural people, it create awareness among rural people of their problems and to built up to trackle key problems in agriculture, health, education, women empowerment, panchayati raj, cooperatives sector etc.

b. **Objectives:** The specific objectives are:

1. To educate the role of some of the major ICT initiatives in rural development
2. To study the impact of ICT efforts on agriculture, health, women empowerment and panchayati raj
3. To point out the impediments for the effective implementation of ICT based project

c. **Data:** The ICT is based on mainly on secondary data. However, it also based partially on the primary data of the Warna Wired Village Project, Waranangar, District Kolhapur (Maharashtra).

History of ICT in India

In India, the path towards technology- induced development especially associated with ICT was given a vent in 1984 under Shri Rajiv Gandhi and the technologist Shri Sam Patroda. A high Power National Task force on Information technology and Software Development was later set up in 1998. This Task Force prepared the blue print for making the adoption of IT as a National Movement by establishing a wide network of empowered taskforce at all government and non-governmental levels. The New Telecommunication Policy 1999 has stipulated targets in terms of establishing a telecom network with a view to achieving tele-density of 7 per cent by 2005 and 15 per cent for a knowledge society.

Information and communication technology is big revolution in Indian farming community regarding the information about agriculture production and market linkages between farmers and mandis. It is great initiative which has been taken by the government in India. The efforts of the Government in the area of ICT has escalated the living standard of Indian farmers and made them ICT-friendly, which has resulted in increased penetration of useful information about crops, soils, climate and cultivation processes.

The concept of knowledge society means that it will incorporate new knowledge will takeover. The role of capital and labour in the economy, further leading to structural and functional changes in family health and education.

ICT based initiatives in India

Initiative	Sponsoring agency	Important features	Major thrust or area or level
Cyber extension project	National Institute of Agricultural and Extension Management (MANAGE), Hyderabad	Village information centres, institutional support to other ICT projects	Post harvest processing, women and child welfare
Agriculture Technology Information Centre (ATICs)	Indian Council of Agricultural Research (ICAR), New Delhi	Call centre, expert systems, researches, farmers linkage	Agronomic practices, technologies inputs

Initiative	Sponsoring agency	Important features	Major thrust or area or level
Agricultural gateway of India	National Academy of Agricultural Research Management (NAARM), Hyderabad	Helpline service through telephone, researches, farmers linkage	Cultivation practices, Plant protection, new technologies
Helpline service	Chandra Shekhar Azad University of Agriculture & Technology, Kanpur	Helpline service through telephone, researches, farmers linkage	Cultivation practices, Plant protection, new technologies
Warna wired village project	NIC, Warna Cooperative Complex, Warnanagar, Dist Kolhapur (Mah.)	Wide Area Network computer booths	Cultivation practices, pest and disease, marketing processing etc.
Gyandoot project	Gram panchayat, Dhar district, Madhya Pradesh	Soochanalayas Portal, Partial recovery, Panchayat community partnership	Agriculture produce, market intelligence, auction rates, land records
Interactive touch screen information kiosk	Dept. for International Development (DFID) Rajiv Gandhi College of Animal Sciences (Pondicherry) and University of Reading (UK)	Information Kiosks text pictures with sound backup	Information about important cattle diseases and management of cattle
Lokmitra project	National Informatics Centre, Gram Panchayat, Hamirpur district (H.P.)	Lokmitra Soochana Kendra, in 25 Panchayat	Classified components, job vacancies, market rates, tenders etc.
E-Choupal	Indian Tobacco Company	e-choupal, crop specific internet	Market price, cultivation, practices, weather etc.
Rural information Kiosks	EID Parry, Tamil Nadu	Farmers managed kiosks portal, public private partnership in applying ICT improvations and content development	Farm practices, farm advisory, maker intelligence, weather etc.

Initiative	Sponsoring agency	Important features	Major thrust or area or level
I-Kisan portal	Nagarjuna group (AP)	Portal in regional language	Agricultural practices, plant protection, animal husbandry, weather etc.
NGOs			
Village knowledge centre	M.S. Swaminathan Research Foundation (MSSRF), Chennai	Knowledge centre, pro-poor, pro-nature, pro-women and community ownership	Agronomic practices, Cattle feed and weather schemes
Computer on wheels	Pingali Rajeshwari entrepreneur, A.P.	Internet on motorbikes, portals, remote area, resource poor	Market, weather-plant protection and animal health
www.doctoranywhere.Com	Tata Council for Community Initiatives (TCCI), Pune (MS)	Telemedicine Service through websites and computer	Medicinal council, medical treatment in PHC

The importance of information in agricultural development and agri-business has encouraged various players to disseminate market information using ICT on different components. Some of the initiatives taken to disseminate information on agricultural and related aspects in the country are depicted in the below Table.

Different ICT initiatives in Indian agriculture

Initiatives	Description
AGRISNET	An infrastructure network existing at block level facilitating agricultural offices, agricultural extension services and agribusiness activities to enhance rural development.
Digital green	The agri. information of local relevance is disseminated through digital video. The system consists of a digital video database prepared for farmers by farmers with the help of experts. The recordings are shown to individuals or small groups using laptops, DVD player, television and to communities through village cable network.
eSagu	eSagu provides personalised expert advice in a timely manner from sowing stage to harvest for small and marginal farmers at their door-step. The farm situation is brought to the expert in the form of digital photographs and text information. The expert advice after analysing the situation is prepared and is delivered to the concerned farmer on the same day or subsequent day.

Initiatives	Description
Warana	The project provides access to a wide range of information including agriculture to the member of the cooperative in local language. It provides information on crops, market prices, employment schemes, educational opportunities, etc. The information is provided through the village information kiosks. The operators of these kiosks are the main linkage between the farmers and the information centre.
IKSL	The relevant information is delivered to the farmers on mobile phones through five voice message in local language. Customized solutions are provided to the farmers through helpline. The farmers can also speak to the experts on specific subject through special 'phone-in' programmes.
Agmarknet	This initiative provides daily market price and arrival information in respect of 300 commodities and 2000 varieties in eight local languages. The wide range of information on prices, arrival and other related aspects like grades, standards, packaging, etc. is collected and disseminated by networking major agricultural produce markets operating in the country.
Pravara	The project aims to connect a hundred villages in Ahmednagar to empower rural population and improve quality of life. The information on government scheme, agricultural marketing, healthcare, education, agro-processing and economic development are disseminated through IT centres established under the project.
iKisan	iKisan is a one-stop solution for farmers in providing information on crops, crop management techniques, fertilizers, pesticides and other related information like market updates and weather forecasts. Earik Single window to improve the access to agricultural information and technology in north-eastern India. It provides expert consultation on production, plant protection and marketing.
Digital Mandi	Digital Mandi is an electronic trading platform for agri-commodities to bring the benefit of ICT to farmers and traders by eliminating geographical barriers and temporal limitation and removing cash crunch through active participation of various financial institutions. Digital Mandi is inspired by the vision of Media Labs Asia sustainable village through culturally appropriate use of new technologies.
Akashganga	The initiative facilitates timely collection of milk, proper payments and generates higher income for dairy farmers. The system includes weighment of milk electronically, fat testing, capturing unique ID by the software and printing of pay slip and payment settlement.
aAQUA	aAQUA is a multilingual online problem solving system that facilitate farmers getting their queries answered by experts. The reply to the queries raised by the farmers to sent in one to three days depending on the nature of the problem.

Initiatives	Description
eKrishi	The communication network established under eKrishi is category to educate farmers, provide real time information on prices, arrivals and issue disaster warning and weather forecast. The aim is to enable farmers to take informed decisions on sale of their produce and bring transparency in the working of the Madhya Pradesh State Agricultural Marketing Board
Mahindara Kisan Mitra	The initiate provide information on daily market prices, weather updates, crop advisories, agri-related news, etc. The information is also available on other sections such as loans, insurance, mandi database, cold storage and warehouses, etc. The farmers can also get motivated and take benefits from the success stories of other fellow farmers reported on the website.
Haryali Kisan Bazar	HKB has set up centre across different states to provide solutions to wide range of problems of farmers under one roof including agri-inputs, financial services, farm-output services and round the clock expert advice. The centres provide information on crops, latest technologies, weather forecast, market prices, categorize services based on the farmer database maintained under the initiatives.
Fisher Friend Mobile Advisory	The information relevant for fishermen is provided in local language through mobile phones. The information covered are wave height, wind speed and director, potential fishing zones, relevant news, government schemes and market price.
KCC	The Kisan Call Centre category telecom infrastructure to provide categorize information on various aspects of agriculture in local language using toll free number 1800-180-1551.
Reuters Market Light	Market Light provides mobile phone based customized information according to the individual farmer's preferences on crops, markets and location. The information in local language in respect of over 440 crops and varieties, more than 1400 markets and 2800 weather locations are available across 13 states through SMS.
e-choupal	An initiative by ITC provides alternative marketing channel, information on weather, agricultural practices, input sales, etc. It is a kiosk located in a village and equipped with computer with internet access managed by trained sanchalak.
e-agri kiosk	An initiative by NABARD and Central Agricultural University. Touch screen kiosk for technology transfer among tribal farmers of Arunachal Pradesh..
MSSRF FFMA	Fisher Friend is a BREW-based application offered on a low cost CDMA handset with a graphic interface, an icon-based menu and programmable shortcut keys. In addition to safety and weather information, fishermen can receive the locations of fishing areas and real time market prices with one-click in their local language.

The above table reveals that a number of approaches have been followed by the service provides ranging from text to voice SMS, digital videos, tele-infrastructure, internet, etc. to get the information delivered to the end users effectively. Sanchalak is a key link to facilitate efficient delivery of information for some of the initiatives. The initiatives discussed above have been categorized on the basis of ownership and delivery mechanism. The table reveals that majority of the initiatives made under private and co-operatives, generally working on self-sustained or revenue model, have relied heavily on Sanchalak, *i.e.*, the facilitator or the linkage between the user and service provider and also mobile phones for effective delivery of the service.

CATEGORISATION OF ICT INITIATIVES IN INDIAN AGRICULTURE

Ownership/ Delivery mechanism	Government	Non-Government	Cooperative/Private/ Consortium
Web-based	AGRISNET, eKrishi, AGMARKNET	–	Pravara, Akashganga, iKisan, aAQUA, Mahindara Kisan Mitra, Haryali Kisan Bazar
Sanchalak (Facilitator between the user and service provider)	–	–	Warana, eSagu, iKisan, e-Choupal
Mobile/ Mixed Approach	KCC, Earik, Digital Mandi, e-Agri Kiosk	Fisher Friend Mobile Advisory, Digital green, MSSRF FFMA	IKSL, Reuters Market Light

The approach with the most effective delivery mechanism will have the highest impact in the Indian condition where majority of the farmers are illiterate or semi-literate. Due to the paucity of research and the relatively shorter period of implementation of such initiatives, it is difficult to identify the approach with maximum impact. Under Indian conditions, the mobile phones being used by a handsome number of rural population and local linkage in the form of Sanchalaks, this seems to be an effective approach. The same has been reflected by private players and co-operatives generally running on revenue models.

The ICT based Model for Information Dissemination

The information is vital for development of agriculture and well being of the rural masses. The fact has been well recognised in the form of a number of initiatives taken to disseminate information on agriculture and related aspects by government, non-government, private and co-operatives. However, the optimum utilisation of the system and the information disseminated will depend on a number of factors like literacy level, understanding of ICT, extent of telecommunication infrastructure, level of awareness of the farmers, information need of the farmers, etc. Any approach incorporating these factors will have maximum impact. Since few research findings are available to support any particular approach, an attempt has been made to identify the important factors for an ideal ICT model for information dissemination considering the above given factors.

i. **Delivery mechanism:** The rural India characterised by poor literacy and poor infrastructure poses the biggest challenge in finding the correct and effective mechanism for delivery of information to the end user. Under such conditions, mobile phones which are used by sizeable rural population can provide an efficient and effective platform for delivering of information. The penetration of mobile in rural areas has increased in the recent past. The tele-density in rural India was reported to have reached 31.18 per cent by the end of year 2010. A study on fisheries sector in Kerala suggests that mobile phone coverage alone led to significant market efficiencies with reduction in waste and the difference in prices across markets. The study also reported an increase of 9 per cent in fishermen's profits and decline of 4 per cent in consumer prices. Another innovation that has been utilised successfully in some of the approaches to deliver information is introduction of Sanchalak - the facilitator. Mostly the Sanchalak is an educated and progressive farmer from the community itself. This linkage already having the faith of the locals can effectively compensate for the low level of literacy and lack of IT exposure of the farmers.

ii. **Two-way process:** Single directional flow of information has been the problem with many of the government extension programmes. An efficient information system should be a two-way process. The content of information to be delivered by the system should be developed after thorough assessment of the information needs of the farmers.

iii. **Integration of various departments:** There are a number of departments operating in isolation to serve the varying and different needs of the farmers. The integration of all these departments will help in economic utilisation of resources and help rapid dissemination of information for its optimum utilisation.

iv. **Window for queries:** Agriculture is a kind of profession where practically farmers come across a new problem every day. The information dissemination system should have provision for interaction with subject matter experts to resolve farmers' specific queries.

v. **Customised information:** Different farmers cultivate different crops with different technology and thus, the system should be in a position to serve the specific requirement of the farming community.

vi. **Integrated approach:** An integrated approach using different mediums like phone, films, digital photos, internet, television, radio, local facilitators, etc. according to the needs and level of understanding of the end user would yield better results.

vii. **User friendly information:** The information should be provided in an easily understandable format preferably in local language.

viii. **Viability of dissemination model:** An ideal model should have financial viability so as to sustain itself over a longer period of time.

There is great transformation in Indian agriculture owing to changes in the economic and trade environment. To cope up with these changes timely, relevant and accurate information to the farmers and other stakeholders will help them take optimum decisions. ICT should play a vital role in the efficient delivery of this information. Several ICT based initiatives have been tried by different players and the same are analysed in the present paper. Based on the analysis review of the different initiatives, an attempt has been made to recommend measures to harness the full potential of ICT as given in next page:

i. Assessment of information needs of the farmers and appropriate mode of reaching them as per local conditions is crucial before developing an ICT Model.

ii. The information dissemination model should be viable and user-friendly so that the initiatives may be sustained in long-run. A string backward and forward linkage should be in place for accurate information collection and its dissemination.

iii. Integration of various agencies under one roof for providing vital information on various components of agriculture so that it will act as a one stop solution for the needs of the farmers.

iv. Introduction of delivery mechanism of information in the case of government initiatives like agmarknet.nic.in is need of the hour so that the information reaches the end user.

v. It is essential to create the requisite ICT infrastructure in rural areas for effective dissemination of information.

vi. Creating awareness among farmers and other stakeholders on the importance of information and its optimum utilisation will help in the development of agriculture and overall well being of the farming community.

ICT: Some prominent initiatives

Let us now compare the ICT initiative in India in terms of Public vs Private/NGOs, their sponsors, features and their major thrust of development:

1. **ICT for agriculture**

   i. **e-choupal:** Indian Tobacco Company (ITC's) International business division is one on India's largest exporters of agricultural commodities has conceived e-choupal launched in June, 2000 as a more efficient supply chain aimed at delivering value to its customers around the world on a sustainable basis. ICT's e-choupal is a unique example of using ITC's for agricultural development, e-choupal has already become the largest initiative among all internet based intervention as in rural India, e-choupal services today reach out to more than 3.5 million farmers growing a range of crops - soyabean, coffee, wheat, rice, pulses, shrimp-in over 31,000 villages through 5,200 Information Kisoks across six states (Madhya Pradesh, Karnataka, Andhra Pradesh, Uttar Pradesh, Maharastra and Rajasthan).

   ii. **I-Kisan Portal (Andhra Pradesh):** The Nagarjuna Fertilizers Company Limited (NFCL) is a agribusiness based at Andhra Pradesh. They are disseminating various farming information to the farmers at various places through ICT centres. To do this job, they have engaged 200 field assistants and 25 technical assistant to serve as contact interface between company and the farmers. They also have Krishi Vigyan Kendra (KVK) at Kakinada (A.P.).

2. **ICT For Livestock:** India has a huge population of 196 million cattle and 80 million buffaloes together producing 84 million tons of milk accounting for about 13 per cent of the world's milk production. The challenges for veterinary institutions to reach millions of livestock owners. Thus, ICT initiative is expected to provide a strong case for effective coverage and dissemination of livestock out reach information to the farming

community. In this context, three e-mail conferencing system, *viz.*, goat-net-cirg.nic.in, livestock-net@cirg.nic.in and fishnet@cirg.nic.in, have been launched by the institute to help information inflow among technologies, farmers, development officers and planners. Under the Animal Health Project funded by Department for International Development (DFID), Rajiv Gandhi College of Veterinary and Animal Science, Pondicherry in collaboration with University of Reading, U.K., has installed a interactive Touch Screen Information Kiosk in one of the rural regions of Pondicherry on a trial basis. It has information on important cattle diseases in addition to management of cattle and methods of acquiring information.

3. **ICT For Health:** Those living in interior Pune village will now be able to avail expert medical council well within their means. The Pune district administration has teamed up with www.doctoranywhere.com and Tata Council for Community Initiatives (TCCI) to launch a telemedicine service from a government Primary Health Centre (PHC). There are at least five to six doctors always present at the headquarters who can respond even if there are no specialists (www.doctoranywhere.com). One another telemedicine project was started in 2000 by Apollo Hospital. The project has already benefited over 3749 patients at 11 remote locations in India. The Indian Space Research Organization (ISRO) and Development and Educational Communication Unit (EDCU) has initiated several telemedicine projects in remote areas. Sankara Nethralaya is a super speciality eye hospital (Chennai) is one of the hospital chosen by ISRO for the pilot project providing telemedicine though satellite connectivity. The mobile tele-opthalmology unit also provides continuing medical education programmes for doctors and create awareness about eye-disease among rural people.

4. **ICT for Pachayati Raj**
   i. **Lokmitra project:** This project was launched in May 2001 in the Hmairpur district of Himanchal Pradesh. Lokmitra Soochana Kendras were set up at 25 centrally located panchayats, National Information Centre (NIC) provided the Lokmitra Software and necessary training to the Soochaks (operators). The services includes classifies complaints, form download, job vacancies, tenders, schemes information, market rates, etc. Some of the other initiative's are GARNDOOT in Dhar District of Madhya Pradesh and SAYTU in Maharashtra are successfully implemented.
   ii. **E-Gram Suvidha:** E-Gram Suvida is Geographical Information System (GIS) based on Management Information System that links the maps of village resolution with the associated data to generate thematic maps based on users queries. An easy to use software, e-gram Suvidha is currently operational in five districts of Madhya Pradesh (www.nic.in)

5. **ICT For women empowerment:** Women being a vital part of Indian economy constitutes about one third of national labour force. More than 90 per cent of rural women in India are unskilled. Proper training to the women and imparting skills through the use of ICT will be quite helpful. The Information Village Project has been implemented by the M.S. Swaminathan Research Foundation (MSSRF) in collaboration with the International Development Research Centre (IDRC). The specific objectives of

this project is to train youth especially women, in rural areas in operating information shops. The info villages provide local-specific information to villagers related to price of agricultural inputs (seeds, fertilizers, pesticides), agricultural output (rice, vegetables) market, healthcare, cattle disease, transport, weather, etc. In local *i.e.*, Tamilnadu language, MSSRF provides this service free of cost and it has won two international awards.

6. **ICT For co-operative development**

   i. **Dairy Co-operative:** National Dairy Development Board, ICT is being used at milk collection centres and in co-operatives to measure butter fat content milk, test the quality of the milk and helps to make payment to the farmers. Amul Co-operative has introduced a number of ICT based operations, which have contributed to efficiency of milk procurement and distribution. The NDDB has also been able to deal with problems of unethical practices in the milk enterprises.

   ii. **Sugar Co-operatives:** The Warna Wired Village Project, covering 70 villages in Kolhapur and Sangli districts of Maharashtra state. The village information centre has provided indepth information about sugarcane cultivation and marketing to the farmers.

Impact of ICT initiatives

The impact of some of these initiative is presented below based on the experiences of villages and farmers:

1. **Reduction in transaction cost:** The transaction cost incurred by the farmers was worked out by computing variables like distance travelled, labour hours involved per visit, cost equivalent of labour hours and travelling cost per visit through ICT enabled services, a farmer could save this travelling time and transaction cost by more than 80-90 per cent. Farmers could place their requirement of inputs to the Sanchalak, who place bulk demand with the ITC. ITC arranged to deliver the products, ensuring quality farm inputs at reduced transaction costs. One-fourth of the respondents contracted under the study reported that they could reduce the cost of procurement of inputs by 4 per cent. Though Gyandoot kiosks, farmers could get the land records in village itself and save transaction costs by 66 per cent.

2. **Time saving:** According to study by UNESCO in 2001, the users in the three knowledge centres of MSSRF, Pondicherry have saved a travel distance of around 100,000 kms, and 4000 hours amounting to imputed value of Rs. 16000. Time is a valuable commodity for the landless labourers and women in the rural areas (particularly during the working seasons) and therefore, time saved means money earned.

3. **Awareness and utilization of entitlements:** Through the village knowledge centre, Panchali, a farm – women living in Kizhoor village pf Pondicherry came to know about the availability of subsidized seeds at PASIC (Pondicherry Agro-Service and Industries Corporation Limited) and procured 40 kg of white ponny Paddy variety seeds at the rate of Rs 10.50 per kg. This timely availability of information saved for both time and money (Rs. 80 subsidy and 40 kg seeds). Access to and control of information by poor women is a step towards empowerment.

4. **Impediments in the use of ICT:** In India, the utilization of ICT in agriculture and rural development is in the take off stage and obviously there are many field problems. About half of the respondents expressed irregular internet connectivity as one of the major constraints, except in the case of initiatives like e-choupal and Computer On Wheels (COW). Farmers faced problems in using the helplines due to very low teledensity in rural Uttar Pradesh. It is widely excepted that ICT's are playing an important role in the development process. But at the same time the most difficulty like inadequate internet/phone connectivity, regular supply of electricity etc.

Suggestions for effective utilization of ICT

Several studies suggested following strategies for effective utilization of ICT projects especially in rural areas:

1. **Content development:** Generic and indigenous knowledge, success stories in local language to meet local needs.
2. **Role of universities and institutions**: Research and development institutions, universities must conduct ICT based project and also conduct Ph.D. and M.Phil. research on this subject.
3. **Integrated approach:** There is need for an integrated approach, based on the real information needs of the potential users and not on mere 'technical fix'.
4. **Enabling environment for ICT:** There is need to create an information friendly environment, characterized by telecommunications reform and information policies, the track of creating such environment falls not to one but all the partners for development.
5. **Holistic perspective:** We need to place ICT in a holistic perspective of information system, as social systems are rooted in a context of people and social structure.
6. **Development of software:** To develop capacity of managers/operators of information services to identify and supply the information needs of users and human capacity (Literacy, numeracy) of beneficiaries to moves use of information.
7. **Bridging the divide:** Strategic development of ICT to address socio-economic dividers: age, gender, community, land holding and geography is essential.
8. **Low cost training institutes:** School and Panchayat Offices should impart ICT training. Students can be effective 'no cost training' NSS students can play key role in this context.
9. **Learning from success stories:** Success stories provide a forum for discussing and evaluating international experience that contribute to lesson-learning constitute best-practices.

The use of ICT in agriculture in general and rural development in particular remains restricted in India. Effective utilization of ICT has the potential to make the rural communities in India prosperous. ICT can definitely play key role in this endeavour if the local Panchayats are encouraged and supported to setup ICT based village development information centres.

Importance of Information and Communication Technology

1. It creates an analytical mind of students that help them study and proper solutions to problems emanating from all related field that employ it as a learning tool.
2. Being an emerging academic field of study, it helps students to be innovative and develop new ways of solving problems scientifically.
3. It makes information storage and retrieval easy.
4. It enhances computer networking globally known today as internet and intranet.
5. It accelerated economic development nationally as it is a virile source of national income for all nations that have fully embraced its usefulness.
6. It creates gainful employment, hence a viable source of livelihood.
7. It makes comprehension of other subjects easy. Virtually all fields of learning are amenable to ICT such as the application of projector for teaching in the classroom.
8. It creates an avenue for the exchange of ideas and inventions among information technology scholars locally and internationally.
9. It is the basis for e-learning and online library. Hence information dissemination is easier than ever.
10. It is pivotal to globalization in all its ramification and the realization of the Millennium development goals as agreed by the United Nations in the year 2000.
11. It is used at various offices for proper documentation of official activities and administration.

Challenges of Information and Communication Technology

Information and Communication Technology is confronted by a number of challenges which is characteristic of an emerging field. The following are some of the challenges:

1. **Expensive ICT materials:** The requisite materials for practical knowledge of ICT especially at the advanced level are costly e.g. Computer, projector, internet machine etc.
2. **Highly technical and practical driven:** It is technical driven. It requires a great deal of logic and analytical reasoning for in-depth understanding and application.
3. **Poor orientation about the concept:** The first challenge in embracing ICT as a field of study for students who have no prior orientation from their foundation education is that of speedy assimilation.
4. **Under development:** ICT is a global concept and in order vape shop have an up-to-date knowledge on the subject there must necessarily be technological development nationally at least well established global information link on the subject. Many nations are yet to attain this.
5. **Universal acceptability as a compulsory field of study:** The subject is yet to be fully embraced by all institutions of learning from the credit globally. Although it is now compulsory at the high school level in some countries. Unless there are solid foundation

and love for the subject, compliance for all may still take a time to attain in line with the global millennial goals.

6. **Hijack by unscrupulous persons:** ICT is being used by some persons for evil purposes such as cybercrime and malicious programs that can cause severe damage to computers and similar gadgets.

Important points

1. India is a developing nation that is making remarkable progress toward development through the informatization strategy. Information is the process through which communication technologies are used as means of furthering socio-economic development.
2. The aims of the task force to formulate the draft of a national policy on informatics to enable India to emerge as information technology super power within the next ten years.
3. Task force June (2001), Steering Committee Members consisted of Dr. A.P.J. Kalam, Dr. R.A. Mashalkar, M.K. Kaw, Dr. K. Venkatasubramaniam, Prof. A. Parthasarthi, 14 members and 3 co-opted members.
4. KIOSKS is internet connectivity for small towns and villages, that is just similar to the STD booths which revolutionized communication by connecting villages to the outside world, through technologies device such as computer, VSAT (Very Small Aperture Terminals), WLL (Wireless Local Loop), LAN (Local Area Network), WAN (Wide Area Network).
5. The farmer-owned Amul Co-operative in Anand has shown that since the 1960s an integrated approach along co-operative lines could enhance production and marketing of milk. It has become a model for most dairy development projects in India.

Traditional Folk Media

Folk media or traditional folk media often appeared as one of the viable ways of solving present problems in development. Local or traditional communication channels such as puppet show, folk drama, story telling, community teahouse, open market and village meetings are available to all and enjoyed by person of different age groups, generally at a very low cost. The Government of India made deliberate efforts to use the traditional folk media for disseminating important themes like national integration, communal harmony, health and family welfare, girl child, prevention of drug abuse etc. a communication system embedded in the culture which existed before the arrival of mass media, and still exists in many parts of the world with a certain degree of continuity, despite changes as a vital most of communication". Traditional communication media like folk songs, drama etc. are highly structured and deeply rooted in the socio-cultural mores of the society. Proverbs, riddles, folktales, myths and folk songs have played an important role in the development of society. Village festivals can serves as occasions for political education, specific development issues can be scripted into folk debates and desirable vale system can be woven into folk drama. Indigenous communication channels are available in many forms such as folk media, indigenous organization and social gatherings, deliberate instruction, record and unstructured channels.

Folk media may constitute a art of the indigenous communication system but there are also other communication media and channels.

Types of Traditional Media

The traditional media can be classified into four major forms.

1. **Folk Theatre Forms**
   - Tamasha (Maharashtra)
   - Powada or pawada (Maharashtra)
   - Dangi tamasha (Gujarat)
   - Dashavatar (South Konkan, Maharashtra)
   - Nautanki (North India)
   - Yakshagana (Karnataka)
   - Jatra (Bengal or Orissa)
   - Bhavai (Gujarat)
   - Therukoothu (Tamilnadu)
   - Burra Katha (Andhra Pradesh)
   - Ram Leela (Uttar Pradesh)
   - Bidesia (Bihar)
   - Jatjatin (Bihar)
2. **Keertana (Harikatha or Harikeertan)**
   i. **Keertana (Maharashtra)**
      - The traditional or Naraditya Harikatha
      - The Warkari Harikatha
      - The Nationalist Harikatha

   ii. **Keertana (Gujarat)**
3. **Folk songs/folk verse/ballads/folk saying and riddles:** There are more than 300 folk musical styles in India of which the more prominent are:
   - The Baul and Bhaitali of West Bengal
   - The Doha and Garba of Gujarat
   - The Chaiti and Kajri of Uttar Pradesh
   - The Kalkli pattus of Kerala
   - The Bihu of Assam
   - The Mnad and Panihari of Rajasthan
   - The Rauf and Chhakri of Kashmir
   - The Dua and Dadaria of Madhya Pradesh

- The Mando and Dhalo of Goa
- The Boli of eastern Punjab
- The Lavani of Maharashtra

4. **Puppetry:** In India four styles of puppetry have proved popular in different parts of the country:
   - Sutradharika
   - Rod puppets
   - Shadow puppets
   - Hand puppets

Form of Indigenous Communication

Folk media: It includes festivals, plays and puppet shows, dance, song, story telling, poetry, debates. They are used primarily for entertainment, but they can also promotes educational values and help in cultural continuity.

Indigenous organization and social gathering: It includes religious groups, village meetings, irrigation associations, mothers clubs and loan association.

Deliberate instruction: Parents teach children, crafts people instruct apprentices, elders guide young people and adolescents undergo initiation rites. Many societies have traditional religious schools.

Record: Many societies keep records in written, carved, painted and memorized forms.

Unintrumental channels: It includes talk at home and at the well, in the field and on the road, in tea house and coffee house, in the chief's house and at the market and wherever people meet and talk. This communication is not organized or orchestrated but spontaneous and informal.

Advantages of using traditional media

1. The traditional media are the closest to the hearts and minds of the masses, and so their appeal is at a personal, intimate level.
2. Their familiar format and content, as also the colloquial dialects used, make for clarity in communication.
3. Cross-cultural communication hurdles are not encountered.
4. Rapport is immediate and direct, the barriers to communication almost non-existent.
5. The traditional media have their own method of communication. Their peculiar style may from the local customs, traditional and various other social practices.
6. These pass the message in the common familiar language of the people.
7. As there is face to face interaction between the conveyer and the recipient, there is a scope for clarifications of doubts and acquisition of full information.
8. These media often appeal to the emotions of members of the audience, who identify themselves with the players.

9. The audience tends to have confidence in the messages because communication takes place in a form they are accustomed to, in their dialect and form people they often knows.
10. Audience are seldom passive listeners or viewers, but tend to participate actively, thus influencing the performers. The Keertan and Alha (of UP) are examples of this.
11. The greatest advantage of the traditional media over the electronic media is their flexibility in accommodating new themes.
12. These media cater to small audience at a time, but these audiences are so completely caught up in the folk forms that the effects on them are at a much deeper level.
13. The traditional art forms satisfy the innate desire for self-expression and also satisfy man's need for moral instruction combined with entertainment.
14. The traditional media have both entertainment and information value.
15. The traditional media are credible and culturally compatible.
16. The traditional media are inexpensive and informal.

Importance of Different Forms of Indigenous Communication

Folk Media

Folk media are intimate with masses, rich in variety, readily available at low cost, relished by different age group and by both sexes and have greater potential for persuasive communication, face to face communication and instant feedback. Folk songs can be both effective and efficient as a communication medium for a society which has a strong oral tradition.

The Indian Government established a Song and Drama Division in 1946. All India Radio also used folk media in its rural broadcasting in the form of a daily programme. Traditional media are important channel of communication. Several traditional methods can be used in secondary school for teaching Home Science. Among these, story telling, folk songs and dances are outstanding to augment in imparting education in an interesting and effective manner. In India has the tremendous varieties of folk singing styles. Folk songs have often been used in India.

Doordarshan has been the pioneering institution to draw upon folk music. The radio, Doordarshan dramas and theatre have integrated various folk theatre forms. Jatra, Tamasha, Nachas, Bhavai, Nautanki, Yakshgan, Therukuthu, Kavi Sammelans and Mushairas are used by various stations of All India Radio. Traditional media have remarkable impact on the rural society because of their acceptable idioms, functional significance, emotional touch and entertainment component. Traditional media are more effective in rural areas because the techniques are simple and easy. Traditional media can more effectively stir the feeling and affect the attitude of illiterate masses.

Traditional Leaders

Traditional leaders play an important role in the lives of rural people and have a great influence in decision making and involving everyone in the communities to so quick spread of messages.

On the basis of above it may be concluded that traditional media may compete for attention with modern media such as radio, television, cassettes or video. A Brazilian agricultural extension service commissioned a popular folk singer to write and perform folk songs which incorporate technical messages about fertilizer use. The Union Bank of India and the Life Insurance Corporation have used the medium of puppetry with great effectiveness in UP to arouse the interest of rural folk in bank savings and LIC policies. The State Bank of India, Pantnagar branch has also used puppetry effectively in Pantnagar Farmer's Fair to arouse the interest of the visitors in bank savings and loans.

Chapter - 14 Information Technology for Agricultural Development in India

- The Information and Communication Technologies can generate new opening to bridge the gap between information haves and information have-nots in the developing countries.
- The task force on 'India as Knowledge Superpower' emphasized the need to harness ICT for community transformation.
- The agriculturally prosperous developing countries like India cannot overlook agriculture in such transformation.
- The emerging ICT have momentous role to perform in agricultural development. There are many possibilities of integration of ICT in agricultural, for the overall agricultural and rural development.
- The role of Information Technology to develop agricultural research, education and extension to improve quality of life in rural area is well established.
- IT can help an average Indian farmer to get relevant information regarding agro-inputs, crop production technologies, agro processing, market support, agro-finance and management of farm agri-business.
- The agricultural extension mechanism is becoming dependent on IT to provide appropriate and location specific technologies for the farmers to furnish timely and proficient advice to the farmers IT can be a best mean not only to develop agricultural extension but also to expand agriculture research and education system.
- The excellent scopes of Internet are waiting to develop agricultural education management through smart exposure of agricultural teachers and educational planners, class rooms, virtual class as well as dropout agricultural learners.
- For agricultural extension management, role of IT can be encouraged for future resource documentation, as methods of extension and linkage between research and extension.
- IT in agricultural research management for textual and non textual documentations and deciding prioritization of research areas needs to be reinforced.
- The crop forecasting, input management, command area management, watershed management, land and water resources development, drinking water potential mapping

precision management, natural disaster management, fishery management, hill area development and post harvest management are the key areas, where Information Technology can play its imperative impact.

Area and scope of information technologies in agriculture and allied sectors

The area and scope of information technologies in agriculture and allied sectors are as:

1. Change in Focal Point

- It is inspiring to note that, in India there is a growing consciousness of the multiple roles that Information Technology (IT) could play in overall growth of the country.
- The politicians and policy makers have also emphasized the significance of exploiting the profit of IT for the overall progress of the country.
- It is a solid view of the Government that if any technology can generate new prospects to link the space between haves and have-nots in society in the current time, it is information communication technology.
- The policy makers have also realized that IT to improve the lives of the two fifth of the population, which lives below the poverty line, the Government must play a catalytic and enabling role.
- Besides the central administration, several State Governments have also committed themselves to make strategies, which intentionally plan as extensively as possible to capture the benefits of IT, including the less privileged segments of society.
- In addition to the Central and State Governments, different civil society organizations are also convinced of the potential role of IT on the socio-economic revolution.
- A number of initiatives towards harnessing IT for rural development and poverty alleviation have been taken by these bodies.
- Even the private corporate sector, perhaps in the context of declining world demand, is increasingly looking towards the domestic market as a source of revenue generation.

2. Various Stakeholders

- Nowadays there are various stake holders-central and state governments, civil society organizations and the private corporate sector.
- Depending upon the character of the actors involved, the information communication technology projects in the country may be broadly separated into two groups. The first group refers to those initiated by the central and state governments essentially concerned with e-governance. The second group refers to the initiatives undertaken by the non government organizations and the commercial sector.
- During the year of 2002, the states like Kerala, Karnataka, Andhra Pradesh, Maharashtra, Bihar, Orissa, Uttar Pradesh, Madhya Pradesh, Gujarat, Haryana, West Bengal, Rajasthan, Punjab, Assam, Nagaland, Delhi and Sikkim have initiated many e-governance to harness the advantages of information technology.

- The inter-state variation in the initiatives for harnessing information technology for development is observed but now all other states have also realized its importance.

3. Regarding the Substance

- Access to information and improved communication is a critical prerequisite for sustainable agricultural development.
- Recent communication technologies when useful to conditions in rural areas can facilitate advanced communication, boost participation, disseminate information and distribute knowledge and skills. It is said that extension through information technology would be the major form of technology dissemination in the near expectations.
- It is experiential that the rural community still has trouble in accessing essential information to make timely decisions. It is necessary that information accessibility is demand driven rather than supply driven.
- The challenge is not only to improve the accessibility of communication technology to the rural population but also to improve its significance to local development. Considering the important need to access to well-timed information and improved communication, this issue is in centre of attention on attempts made in different countries to transfer information to the rural population.
- There is a concern that the gap between the information rich and information poor is getting wider.
- New ITs are creating to solve problems of rural poverty, inequality and giving an opportunity to link the gap between information-rich and information-poor and support sustainable development in rural and agricultural communities.
- However remote rural communities still lack basic communication infrastructure.
- The challenge is not only to improve the accessibility of communication technology to the rural population but also to improve the relevance of information to local development.

4. IT in Agricultural Education Management

- Since independence, the higher education system has undergone many changes. The number of universities has gone up from 19 in 1947 to 275 in the year 2000. At the same time the number of colleges have increased from 591 to over 10,000 and students strength from 0.2 million to about 7.0 million. Today with over 3, 00,000 teachers, the Indian higher education system is the second largest in the world.
- Considering this scenario of agricultural education the enrolment of students in agriculture remained only 1.5% in agriculture and allied fields. However the percentage of students studying for professional degrees is extremely low.
- With this background in view, we can be able to understand, how difficult it is to face the challenges of higher education in 21st Century which is being dominated by the Information Technology. This clearly indicates that all out efforts need to be made in

higher education especially in the field of science and technology to harness the youth potential of rural India.

- This could be only possible through increasing the accessibility of higher education through IT and innovative mode of education such as Distance Education or Correspondence Course System. At this stage intervention of IT is required.

5. IT for Agricultural Teachers and Educational Planners

- Successful integration of IT in Agricultural Education it needs to empower teachers of agricultural schools, colleges and universities and also administrators and educational planners. For this purpose at the elementary stage, it required to arrange training of these personnel in fundamentals of Computers and then gradually introduce the advanced modules for Computer applications.
- Teachers need to be trained to use computer to prepare their lesson plan, presentations 'on power point, photo scanning and use of LCD. The educational planners and administrators should be trained to prepare Annual Budget Plan, for teaching aids, resource person and material expenditure, infrastructure budget requirements, time tables to monitor and scheduling the teaching resources, to build up and maintain comprehensive students records and students files etc. Teacher's role has been changed in IT based agricultural education.
- The Teachers of IT based agricultural education need to be trained to prepare for "Virtual Class Room". For this, they should prepare slides, record their sound of lecture on Computer itself and this recorded lecture is being attended by his learners in 'Virtual Class'.
- The module courses will have to be designed for virtual Class Studies and 'Network Based Education' or 'Information and Communication Technology (ICT). IT along with Internet Expertise is boon to the Distance mode of education. Under Open University or School Education pattern, the learners do attend classes by visiting their respective study centres and attend lectures of their teacher counselors.
- However, with the introduction of IT the Learner need not come to institutes and study centre but they can attend his virtual class at his home. For this purpose teachers are required to train to prepare 'Training Capsule' for virtual class through which the students or learners can access on their internet facilities.
- The teachers will have to be trained in E-mail, chatting, surfing, teleconferencing, video conferencing and all latest communication technologies.

6. IT in Agricultural Learners in Class Rooms

- The learners or students in Agricultural School and Colleges need to be well acquainted with 'understanding the lessons projected on the LCD Screen, using power point presentation.
- Until now the learners of classrooms are in habits of listening lectures with the help of overhead projectors (OHP). Sometimes video films, online internet presentations can also be shown on screen with LCD.

- The use of CD ROM, on specialized topics on agriculture can also be displayed on computer monitors.

7. IT for Agricultural Learners in Virtual Classes

- With the advent of Internet, Technology, the learners can attend 'Virtual Class' on the monitor of their computer at their homes or workplace.
- The Learners will have to be trained in using Internet Technology *i.e.* in this case how to log on, do searching and save the information on computer itself.
- After attending 'Virtual Class' on Computer, one can appear for 'ON LINE' examination test wherein the student will be required to type answer on computer and he will know his evaluation report on screen immediately.
- Virtual classes are stepping stones for building future Virtual Colleges and Virtual Universities. The Virtual Colleges and Virtual Universities are equally useful for both thinly and thickly populated countries.

8. New IT Dimensions of Agricultural Education in India

- To prepare the agricultural graduates capable to meet the challenges of the new millennium, they should be given course on International Agriculture, WTO, Trade Related Intellectual Property Rights (TRIPs),
- Global Conventions on Climate, Biodiversity and Desertification, Computer Technology, Patent and Trade Literacy, International Standards. For this IT can play important role.
- Introduction to Computers, Application of Software, Data Base Management Systems, PowerPoint, Drawing Software, Computer Programming, Multimedia, Internet, role of TV and Radio in ICT should be a part of their course curriculum.

9. IT for Agricultural Empowerment Dimension

- There are a large number of dropouts in rural schools and a major segment of the rural population is still unreachable.
- Every State Agriculture Department should establish independent cells for Distance Education for training these dropouts and rural youth for imparting short term courses in newer skills in agriculture and allied sciences to improve farming. For this IT can play important role.

10. IT to Link all Agricultural Colleges of India

- Nowadays there is no any direct link between all the agricultural colleges of India.
- IT can be an effective mean to link all the agricultural colleges of India. In academic field if we have good contacts with other institutions we can be able to know their academic development and compare our work with other them.

- This kind of contacts gives us opportunity to make necessary changes in out style of work for desired result.
- IT facility makes person cosmopolitan in nature. Exchange of useful information like collection of question papers, recent trend, information on seminar, symposium, workshop, training and any other academic developmental activity can be known easily through the IT.
- It should be the foremost duty of Indian agricultural educational planners to introduce IT in all the agricultural colleges of India and link them with each other to make best use the libraries of agricultural colleges and faculties.
- Some basic reforms required for this is training to the library personnel of the agricultural universities and colleges in latest computerization process of libraries with newly introduced IT.
- The establishment of databank in libraries facilitate will be useful not only to the researchers and teachers but also very useful for students as a future researchers.
- The human resource management for agricultural education has also assumed an extra ordinary significance as the IT involvement has supplemented a new breadth of virtual classrooms, accessibility of information through Internet.

11. IT in Agricultural Research Management

- The major contribution of agricultural research in India has been reflected in various agricultural revolutions during the post independence period.
- The result of agricultural research boosted the food production and we could see the Green, White, Blue and Yellow revolutions in the fields of Cereal crops (wheat), Milk, Fisheries and the Oil Seeds witnessing the Golden Revolution of horticulture crop production.
- However with the advent of new emerging agricultural technologies there was a change in focus from increased production to increased efficiency.
- The new areas of concerns for agricultural research included the sustainability in agriculture, food security and demand driven research than merely the supply driven.
- For this purpose, the findings of laboratory research need to reach the unreached. Further for this dissemination a careful documentation methodology and proper communication media will play a significant role.
- In this sphere, the Information Technology can be fully utilized for proper transfer of technology to the farming community and also those living in remote areas of villages. However there is still scope for tapping and harnessing all available resources in areas of application of Information Technology.
- It is worthwhile to strengthen the role of IT in agricultural research management. IT can be best mean for research documentation, experiments and analysis of results presentations.

- In the changing scenario under Indian and global context affect the entire process of agricultural research, especially the identification of thrust areas of research.
- The skill to distinguish between what is urgent and what is important will hold key to the success in deciding priorities. Such skill can be acquired by IT.

12. IT for Research Documentations

- Textual documents – To present information in the form of Written text e.g. books, periodicals, catalogues, statistical compendia, trade publications, patents, etc..
- Non textual documents - e.g. maps, plans, graphs, diagrams, posters, paintings, photographs, slides, sound tapes, films, videotapes, artistic monuments and magnetic documents for computer processing IT can be the best mean in agricultural research management.

13. IT in Prioritization of Research

- The changing scenario under Indian and global context affects the entire process of agricultural research, especially the identification of thrust areas of research.
- The skill to distinguish between what is urgent and what is important will hold key to the success in deciding priorities. Such skill can be acquired by IT.

14. IT in Research Communication

- The benefits of Internet connectivity can be utilized for better collaboration amongst scientists for exchange of their views.

15. IT in Agricultural Extension Management

- The present age has been rightly termed as an "information age". People want adequate and authentic information as early as possible.
- Farmers as human beings are also anxious and become more desirous with the advancement in science and technology to know what is happening in the field of agriculture.
- Farmers have enthusiasm to obtain knowledge, particularly in the field of modern agriculture to become psychologically strong and conducive with necessary capacities to adopt modern methods of agriculture.
- In India, it is very difficult to contact each and every farmer in limited time to communicate latest agricultural technology.
- To diminish this difficulty, various mass media are certainly most effective avenues to convey information to the broad means of people, particularly to the huge illiterate segment of the farmers.
- In India among various media, radio, television, literature and newspapers are certainly most utilized by the extension workers to transfer agricultural technology to the huge illiterate and literate segments of the rural.

- A great transformation in agricultural extension approach in dissemination of knowledge.
- This revolutionize is due to more expansion in farming system fairly than the earlier accent of yield improvement.
- Even at previous junctures of agriculture extension growth, the style was information provide oriented to a certain extent than farmers' demands driven.
- The advancement made in information technology is so fast that every areas of livelihood have to be well organized to strap up such technology.
- It is experienced that generally the benefits of information technology have restricted primarily to the urban areas.
- This is only due to lack of understanding about the new Information Technology. However as a result of new approach of strengthening the communication and training centres like Agricultural Science Centres and Farmers Training Centres, the whole agriculture extension structure has been reinforced again.
- With a view to understanding the impact of IT in transfer of agricultural technology, the resource documentation and its application play central role as the agriculture extension approaches are shifting gradually.

16. Planning for Future Resource Documentation

- The production of CD ROM on special topics can be the best mean for future resources documentations.

17. IT in Methods of E-Extension

- This a new term coined for electronic extension approach, which is otherwise can be called as I.T. oriented Extension.

18. For the Linkage between Research, Extension and IT 's Role can be Encouraged

- The network between different agencies like Agricultural Science Centres (Known as Krishi Vigyan Kendra), Farmers Training Centres, Agricultural Technology Management Agency and Information Shops needs to be developed for useful linkage and proper utilization of available resources.
- The human resources will have to be trained in usage of IT Tools and all infrastructure facilities required for strengthening the Agricultural extension System and Services.

19. IT in Agro-based Rural Development

- It is assumed that 60 to 85% of household consumption belongs to agricultural products so agriculture plays important role in industrial development, it provides raw materials to industries like cotton textiles, jute, sugar, tobacco, edible and non edible oils, leather, plantation industries etc. The food processing industries is also dependent on agriculture.

- Agro based materials are exported in European and Gulf countries by India. In all such agro -based industries, role of IT needs to be improved.
- IT Tools are very useful in creating effective linkages in agro based industry activities. These linkages are concerning dissemination of useful information.
- Linkages of the producers can be with State Federations and National Federation and Board, Finance Corporation. Advertisement is best way to add value of products.
- This market again can be very well established with available database of product wise information on products with comprising data of competing nations of the world. IT can help in this direction.

20. IT in Agricultural Production

- The IT Approach for commercial crops, horticultural crops or floriculture have to focus on Integrated System may be for plant nutrition or plant protection.
- The well established Integrated Plant Nutrition Approach and Management and Integrated Pest Management (IPM) need to be strengthened with the help of IT Tools.
- The Post Production Technology (PPT) needs to be utilized properly.
- The end user, beneficiaries and all concerned especially with export of agricultural produce need to be trained to access the Internet facilities available as one of the most useful IT Tools of the computer era.

21. IT as Geographical Information System (GIS) in Agriculture

- The use of IT through GIS is very encouraging in India.
- The important areas like Crop forecasting (procurement policy, crop insurance, relief measure), Cropping System (input management: fertilizer, Crop Diversification, intensification, degradation measures, sustainability measures), Command Area Management, Watershed Management, Land and Water Resources Development, Drinking Water Potential Mapping Precision, Natural Disaster Management (flood, drought), Fishery (inland, Marine), Hill Area Agriculture Development Management, Post Harvest Management and Precision Farming can be reinforced with the help of Information technology in India.

Indian Agricultural Web Sites

- www.ycmou.com/agri
- www.khetiwadi.com
- www.kisan.net
- www.krishiworld.com
- www.nic.in/agrico
- www.pravara.com

- www.agrinto.com
- www.nabard.org
- www.agriwatch.com
- www.ciks.org/agri.html
- www.agri.mah.nic.in

Scope of rural Internet

Enormous benefits await rural communities and agricultural organizations when communication improves between the non-governmental organizations, government services, private sector and educational institutes that support rural and agricultural development. By sharing information about their activities in the fields of agriculture, rural development, forestry, fisheries, health, nutrition and education, these agencies can better serve rural people and farmers. At the same time, rural communities and agricultural organizations can benefit equally from improved vertical channels of communication that enable rural extension agencies and farmers to communicate with decision-makers and others concerned with development.

An integrated approach provides for vertical communication by establishing rural Internet access sites, and by enhancing horizontal communication between such entities as agricultural development organizations, agricultural input and equipment suppliers, government extension services, rural development organizations, health care agencies, and agricultural research and documentation centres.

Improved horizontal communication can also include existing media services that serve rural stakeholders. For example, throughout the developing world, rural radio and, increasingly, television broadcast services, are important information delivery mechanisms. Their services improve significantly through the exchange of information and news by way of the Internet. Internet services, in conjunction with existing and more widely used communication media such as rural radio, will enable the broadest enhancement of information and communication resources for rural people. Improved horizontal communication and improved information resources can improve the quality of the decisions and interventions that impact upon rural people.

Issue of IT based Extension in India

The information for development from developing countries perspectives and the database creation of local knowledge and practices according to their demand is little available or not at all. More than the access to technology, content creation seems to be a very difficult task. Take the case of indigenous knowledge and traditional agricultural practices in a country like India, which is valuable. Eminent people have been emphasizing the need to document such practices to help sustainable development for a better world. Unless a specific content creation effort is going to take place, there is a greater risk of this knowledge being endangered.

The developing nations will have to borrow the technology and not necessarily all the content. In order to arrive at a reasonably clear picture of the country of existing communication habits and channels in rural areas, especially among the poorer households, from an analysis of the available data, certain trends emerged. For example, of the total households, the telephone density is as low as phones per household. Reach of electronic media, especially television, is reasonably high when one considers the prevalence of poverty in the villages of India.

Establishment of information centres in villages

To provide information to local rural people through satellite based communication technology according to their needs and demands, Information centre at village level can be established by involving local people in choosing the actual location of centre, providing rent free space and electricity and recruiting volunteers. Information centres in isolated rural communities with Pentium personal computers, printers, a modem, a fax machine, Internet and electronic networks, access to databases, libraries and a consultant can make information accessible to a wider audience. Such centres not only facilitate single-point access to external information services e.g. government marketing and price information or to global information through the WWW, but also help in organization of virtual village-to-village meetings and tele-training events thus facilitating local sharing of information. Each centre should contain data on agriculture, health-related information especially for the rural farmers, women and children.

Application of IT based agricultural communication in India

There are cases of application of information and communication technologies in extension that have made a difference in the delivery of extension services in rural India. Some of these include the Warana Wired Village Project in Maharashtra; Milk collection in dairy co-operatives (National Dairy Development Board); Information Villages Project (MS Swaminathan Research Foundation-International Development Research Centre); Information Technology Application for Indian Rural Postal System (CMC Limited, Hyderabad); Knowledge Network for grassroots innovations (IIM, Ahmedabad); Application of Satellite Communication for Training Field Workers and Extension Workers in Rural Areas (ISRO); Computerisation of Mandal Revenue Offices (MROs) and Computer Aided Administration of Revenue Department in Andhra Pradesh

Warana Wired Village Project

Warana Nagar, a cluster of 70 villages in Maharashtra is a central eye of the "Wired Villages" project. In 1960, a visionary like Tahasaheb Kore propagated the idea of cooperatives in Warana Nagar, as a method of achieving socio-economic development. He showed how this could bring all the farmers together; to share information, increase productivity and profits. Thus was born the "Warna Nagar Co-operative Society". The society has a Chairman and a Board of Members and is free from political influence and society members are free to elect the board members. There are about eight sub co-operative bodies, working under this main society *viz.*; Warna Dairy Development Society, Warna Cooperative Bank, Warna Foods, Warna Women's Co-operative society etc. Sugarcane is major crop of this area and most of the sugar production of the two districts Kolhapur and Sangli is processed at this Society. From each village 200 - 300 farmers are registered as society members. The "Wired Village" project was initiated by Mr. Vinay Kore, the son of Mr. Tahasaheb Kore and the present Chairman of the Warna Co-operative Society two years ago and actual implementation began in April 1998. The Project has been jointly implemented by GOI through National Informatics Centre (NIC),

Government of Maharashtra and Warana Co-operative Society with the share of financial support being in the ratio of 50:40:10. The manpower and maintenance cost is borne by the Warna Co-operative Society itself. The project area is a cluster of 70 villages consisting of 46 villages from Kolhapur and 24 villages from Sangli districts of Maharashtra. This project has been initiated to serve the information needs on different crop cultivation practices of major crops, sugarcane cultivation practices, pest and disease control, marketing information, dairy and sugarcane processing information etc. to the farmers, right up to their village level. NIC, Pune was involved in setting-up the hardware and software and NIC, Delhi established connectivity of WAN links such as VSAT and dial-up connections. The software required for the system such as web page designing, database designing and client based applications used by the farmers such as dairy; sugarcane information systems had been developed by the NIC, Pune.

The Computer Booths are serving as information centres for the farmers in their villages. The computer booth is operator by the booth operator and he is the main linkage between the farmers and information gateway centre. The information sought relates to crops cultivation practices, land development, pesticides, diseases control details, marketing details, bills payments positions of sugarcane and dairy etc. Currently forty-six computer booths are functioning in Kolhapur and meeting the information needs of the farmers. In remaining 24 villages of Sangli district, computer booths and hardware was setup, and are waiting to link to Central Server Station. Apart from information retrieval, there are two client-based applications, to serve the farmers needs. They are (1) Dairy Information System (2) Sugarcane Information System.

In Dairy Information System, the information on all the farmers, who are part of the dairy system, is maintained. Other details available to members of the dairy cooperatives include the quantity of milk supplied by each farmer, fat content, their billing information and credit details etc. This information is maintained and updated at the central database on daily basis.

In Sugarcane Information System, information on shareholders is maintained to provide guideline for sugar cane crop to about 200-350 shareholders in each village. This system maintains the details of the cultivation schedule, quantity harvested and supplied to the society, deductions effected by the Society towards credit, net income due to the farmers is available with respect to each shareholder. Every village is also linked with the Directorate of Marketing in Pune, which facilitates farmers in getting information on rates of vegetables, fruits and other crops.

The Information Villages Project

This project is implemented by the MS Swaminathan Research Foundation in collaboration with International Development Research Centre (IDRC), aimed at bringing the benefits of modern information and communication technologies to rural families in Pondicherry.

The prime objective of this project is to assess the impact of ICT in promotion of sustainable agriculture and rural development and document their role in supporting the process of knowledge empowerment of rural families. Specific objectives of the project are:

1. To set up six Village Information Shops that enable rural families to access modern information and communication technologies .
2. To train educated youth especially women, in rural areas in operating Information Shops.

3. To train rural youth in the organisation and maintenance of a system that generates locally relevant information from generic information.
4. To maintain, update and disseminate information on entitlements to rural families using a blend of modern and existing channels of communication.
5. To measure the impact of information shops and ICT through surveys, participatory rural appraisal and other methods.
6. To build models of information dissemination and exchange in rural areas that use advanced information and communication technologies.

The Project is located in the Pondicherry region in South India. A Value Addition Centre (VAC) has been established at Villianur and is functional since February 1998. It acts as the hub of the communication network in the project. Four Village Information Shops have been set up at Kizhur, Mangalam and Embalam, and Veerampatinam. An office in Villianur, serves as a Value Addition Centre. Villianur is a market centre for many hamlets that surround it and is also an administrative node and a road junction. Villianur has been selected based on the access of rural families to infrastructure and markets. Project staff at VAC scans the WWW for useful contacts and technology. Data gathering and value addition to data are carried out and information is transformed to suit local queries or needs. This is also an exchange point for a variety of locale specific information on health, transport, public events, subsidies, prices etc. Information on developmental programmes (entitlements, credit, inputs etc.) and markets is also maintained.

Indian Space Research Organization's Project

One-way video, two way audio teleconferencing interactive networks have been used for education and training by Indian Space Research Organization. The major application of the network in rural development was for training extension staff from various departments of the state governments. In addition, a large number of women, Panchayati Raj elected officials, primary school teachers and child development workers spread over large distances have been trained.

Satellite Krushi Gosthi

Like all other State Agricultural Universities, Gujarat Agricultural University also performs triple functions of teaching, research and extension education. The research generates technologies, which can be utilized by farmers and rural people. The present system of the transfer of technology from Gujarat Agricultural University (GAU) to extension functionaries of the development of the State and in turn to the ultimate users consumes considerable time. Looking to this reality the GAU has prepared a major plan under the name of "GAU Satellite Krushi Gosthi" to apply modern tools like satellite linkage for agriculture sector. The GAU is the first in all SAUs, where such kind of facility has been installed. The GAU satellite Krushi Gosthi for transfer of technology can reduce the time lag to a considerable extent the system helps for large area coverage as well as noticeably reduces the distortion in message transfer. Such facility provides facility for two-way conversation. It helps farmers to get on the spot solutions of their questions and queries regarding the live programmes while watching it at the classroom end. This facility makes possible to keep a live contact between the scientists of the university and the farmers of the state.

A satellite based distance interactive education system normally consists three elements, first TV studio from where scientists deliver the talk through live programme, second a number of remote classrooms or Direct Reception Centres (DRSs), with the facility of TV set and STD telephone, from where farmers can watch the live programme on TV sets and third satellite linkage to transmit live programme given by scientists from the TV studio to farmers at DRSs. The Gujarat SATCOM Network has full capability for one-way video and two way audio. RESCO has established SATCOM Network consisting of TV studio at the capital city of Gujarat. The video and audio from TV studio are digitally transmitted to the classroom ends (DRSs). The return audio at classroom (DRSs) is available through STD lines. This facility is used to keep a live contact of the scientists of the university delivering a talk from the studio, with those farmers who are watching live programme at the classroom ends. At present more than hundred Direct Reception Stations (DRSs) to receive transmission are already established throughout the state with the collaboration of different departments of Government and NGOs. GAUSATKRU has vital linkage with them. This system helps farmers to receive information regarding inputs as well as markets. This latest satellite based communication facility is also useful to the students of the university to interact with the dignitaries or experts of agriculture field. To reduce time lag to a considerable extent, such type facility can also be installed in other State Agricultural Universities. Such facility needs to be strengthened at village level with the collaboration of NGOs, schools, co-operatives and Government organizations.

Government of Andhra Pradesh's effort

Satellite based Information and communication technologies are an important part of the Government of Andhra Pradesh's efforts to improve the efficiency of its administrative offices. AP is the first state in India to design a statewide computerization program that will be used in rural areas, at the administrative unit above the village-level panchayat. There are 1124 mandals in the state. The first software application, is the issuance of certificates pertaining to land holdings, caste, nativity and income across a common counter, without the current delay of 15 to 20-days. The AP State Wide Area Network (APSWAN), aims to link the state government's Secretariat with 23 District Headquarters, serving as the backbone for "multi-services" (voice, video, and data) that would be used for improved co-ordination between state headquarters and district offices in managing various regulatory, developmental and hazard mitigation programs of the state government. Mandals will be served by this two-way communication, and electronic commerce applications will be developed. The AP Value Added Network Services project hopes to deliver a variety of public services through a large network of information kiosks. The Computer-aided Administration of Registration Department (CARD), a project of A.P. aims to introduce a transparent system of property valuation, which is easily accessible to citizens.

MANAGE's efforts

The National Institute of Agricultural Extension Management, MANAGE, Hyderabed, has taken-up a number of "Cyber Extension" initiatives, across the country. District level Web Sites are being hosted, Information Kiosks are being established at block/ Mandal and village levels and technical and other need based information is being collected, digitized and hosted on the Internet.

Extension on the Web

The web was first used to deliver agricultural content to rural US farmers almost ten years ago. The problem is that this use of the web remains at an early stage around the world. It is likely that the Web will expand gradually to many more rural areas of the world because farmers demand it, needed technologies are getting better and economies of scale are lowering costs.

Extension services and other content providers face high costs to develop and maintain web resources. Content might be kept offline to protect print sales or it may be posted only at a fee. When content is online and free, its posting is often uncoordinated with other providers, making it harder to find and utilize. The creation of ideal websites by key policy makers for extension through involving its real users in planning, implementing and sustaining this modern satellite based extension system.

Production of Ideal Web for extension

The WWW is often referred to as the 'Web' and provides users with easy access to information resources and services on the Internet. It supports a variety of media formats: text, graphics, images, animations, sound, and video. The Web can be viewed as a multimedia subset of the Internet. Web pages contain multimedia information, hyperlinks and scripts for non-linear navigation. The web pages not only provide users with access to information but also are increasingly being used for communication and collaboration. The 'Web Browser' is a software application that enables users to access resources on the Internet. The first generation of browsers was limited in their functionality and supported only text files. Later generations of browsers are very sophisticated and support multimedia capabilities. They are designed with a graphical interface and are very easy to use.

Creation of web site

- **Web site:** All your pages, images and other files make up your "Web site".
- **Home page:** The first page you want people to see in your set of pages is called the "home page." Other pages are just "pages".
- **Web Pages:** The documents you see on the Web are called "pages". They contain text, images, and "links" that let you jump to other pages.
- **URL:** The strings of characters (like http://www.gauanand.com/) you keep seeing in advertisements are locations on the Web called Uniform Resource Locators or "URLs". You can pronounce this as "you-are-els" or "earls".

To accelerate use Internet facility by the farmers, agencies involved in satellite-based extension should also create effective Web sites for the farmers.

There are some important points are recommended for small web development:

1. **Conduct Informal Strategic Planning:** Follow these basic steps of group problem-solving to develop plan: Set goals, benchmark, brainstorm, prioritize ideas and develop an action plan. Include farmers, agribusiness and Govt. officials in the process. Post and update plan to guide development and show progress. Acknowledge contributions to the site by staff and stakeholders.

2. **Prototype and Revise Sites:** Favour rapid prototyping and formative evaluation of Web sites rather than long, drawn-out stages of design and evaluation. Don't be afraid to post preliminary content and Web page designs. Unlike the print world, Web 'publishing' is a continuous process of creating and refining documents and page designs. Be flexible and let the process evolve. Typically short fact sheets and brochures go online first, then longer publications. Eventually it can develop decision tools, databases and other interactive features.
3. **Identify Content:** Practical coverage of specific crops and growing areas is a must. Examples of high demand content include: Climate and weather, variety testing, pest management, soil treatments, current market prices and government regulations and programs. List internal content you have in-hand and prioritize it for posting. List and evaluate available external content to which to link.
4. **Create an Online Crop Bulletin:** Bulletins of timely crop and pest information during the growing season have proved popular with Illinois farmers it will serve as an entry point to attract users to your full range of online content.
5. **Post Handbooks for Farmers:** For example, the online Illinois Agronomy Handbook offers full text chapters, interactive tools and databases. Calculators for nitrogen rate, limestone, seed rate, replant decisions, stand counting and yield estimation allow users to plug in values and to save results for later use. Large databases with climate and soil information provide inputs to calculators.
6. **Consider Outsourcing Initial Work:** There is no lack of talented Web development experts to maintain a Web server, design sites, post content and create interactive features. Many charge reasonable fees and complete work promptly. They do not need to be on-site; telecommuting does not work for some professions but it works for Web production.
7. **Start a Web Production Team:** When you form your own team, hire individuals with multiple skill sets that "can do it all." Or nearly so, programming and design skills do not often mix in the same person. Keeping your core staff small saves funding and helps avoids the bureaucracy created with larger production teams.
8. **Plan for Convergence of Media and Reusability of Content:** Capture synergies by delivering the same content over multiple media. This strategy is central to the much-discussed convergence of traditional media (Snail mail, radio, print and TV) with new media (e-mail, CD-ROMs and the Web).
9. **Adopt Standards of the World Wide Web Consortium ("W3"):** W3.org oversees the Web. Following W3 standards makes your Web site as accessible and functional as possible. For example, their meta-data standards help insure that your content is highly searchable. A caveat is that page design for display in Internet Explorer is a de facto 'standard' given their enormous market share among Web users. Design for compatibility with Netscape is important when feasible. Design for accessibility is important, sometimes even the law. Enter your URL at bobby.org to receive a free report on the accessibility of your site.
10. **Make Web Server Policies:** Adopt a strategy to manage staff log-in and passwords to the Web server. Decide which software is best for your staff to create web documents.

For example, FrontPage is easy to use but requires special web server extensions and creates non-standard source code. Dream weaver is more powerful and follows W3 standards but novices often find it difficult. Plan a URL scheme before your URLs multiply and get out of hand. Insure their stability, consistency, elegance and intuitiveness. Gradually set standards for Web design including official logos, footers and acknowledgements. Consider use of free open source software like Linux and Apache. Monitor server security issues such as hackers and viruses and make sure your server software is routinely updated with security 'patches.' Provide search capability to your web site by using the free search capability offered by google.com and other search engines.

11. **Market your site:** Submit your site to major portals like Yahoo and to portals specific to Agriculture. Add metaname entries to your source code to attract search engines. Since 'preferences' of search engines vary, review the site http://www.searchenginewatch.com to understand how to tailor meta-names for various search engines. Check by typing keywords from your site into different search engines to check your visibility.
12. **Create both National and Regional Portals:** Begin with a national portal but plan and create local portals. The national portal aggregates content into a comprehensive knowledge base of extension information. For example, the Agricultural Gateway to India is a national portal that includes listings of regional and local portals. Several models of community Internet projects provide guidance for planning local portals. For example, perhaps the earliest effort was the M.S. Swaminathan Research Foundation's village knowledge centres in Pondicherry, India. The TARAhaat portal and Babhaleshwar KVK web sites in India also offered early examples of community portals. Newer projects on the scene include Drishtree and n-Logue.
13. **Site Design Considerations:** Ease of navigation, quick download and good search ability of your site are key design goals. A comprehensive, regularly updated site index is a must. Design for a lowest common denominator equipment but support higher end users. Assign someone to maintain top-level pages to keep the site dynamic. Have a "What's New" section. Create multi-lingual versions of your most valuable content.
14. **Try Web Delivery of Traditional Programs:** Traditional face-to-face programs can be supplemented or replaced with Web-based versions. As supplements they can serve as a primer and follow-up to normal face-to-face programs. As replacements, they can deliver content by slides annotated with voice and by more production-intensive streaming media. Consider use of "virtual classroom" software that offers easy creation and delivery of online programs with presentations, discussions and quizzes.
15. **Require Extension Staff to Produce and Post Content:** Web production is becoming as common in the everyday workflow as phone calls, word processing and PowerPoint. Content specialists do not need to be Web specialists but they do need to be able to routinely post content. Content saved as a print document can be easily saved as HTML. Provide a standard design template that allows them to create, post and maintain content without worrying about design. Content specialists can even record

multimedia presentations from their desktop. Multimedia created on the desktop won't win awards for technical production but it might for delivering timely content.

16. **Require Extension Staff to Utilize Internet Communications:** Staff should routinely utilize e-mail, messaging and even voice/video over IP. It complements your Web-based programs and captures efficiencies. Communications will be speedier, electronic archives can be maintained and ties to clients and stakeholders strengthened.

17. **Train Extension Staff to use the Web:** Deliver training on key software over the Web whenever possible. Use existing online tutorials such as ones produced by the Land Grant Training Alliance. Provide a virtual help desk for staff (and site visitors) that receives and replies to instant messages and e-mail requests for help on software issues. Always archive the replies into lasting FAQs.

Considering the basic preferences of extension Web users:

An effective use of extension is possible; if it is created looking to the expectations, needs and interest of its real users. Some basic preferences of extension Web users are explained.

1. **Data Visualization:** The use of visualization technologies like Global Information Systems needs to be strengthened, GIS has begun on some extension Web sites. Extensive databases of GIS data for a region are compiled from sources like water, geology, and natural history surveys. Before these Extension services combined such data and made it Web-accessible, it was not possible to view data layers created by multiple units of a state government in one place.

2. **Simulation:** Simple Web-based simulation tools are a reality but it will take longer to create robust systems that allow farmers to run complex simulations. Farmers would enter their own data into a simulation engine in order to model crop growth over the growing season based on inputs like climate, soil types, seed type, and soil treatments. Projections of the bottom-line of alternative planting scenarios would also be calculated. Impact on natural resource systems is being worked on too, allowing farmers to assess impacts of soil treatments and other inputs on watersheds.

3. **Ubiquitous Access:** Farmers would be able to access your documents, tools and databases anytime and anywhere. Such capability has been prototyped in the delivery of climate information and agronomy calculators to cell phones and PDA units like the Palm and Pocket PC.

4. **Digital Library Systems:** On-line documents grew rapidly at many extension web sites in the 1990's, but these were often posted haphazardly, without meta-data and on different servers, or hidden from search engines in databases. Extension services are now taking steps to create digital libraries of their content. It is wise to consider needed document management strategies such as meta-data and archiving early in your efforts. Distributed digital libraries are a key goal for the coming decade, allowing farmers to transparently search for resources across multiple extension sites, and effectively creating a world-wide library of food, agricultural and natural resource information.

5. **Integrated Farm Management Systems:** Web-based decision aids for farmers have multiplied in recent years and now they are being integrated into farm management

systems. Users gain the value of one-stop shopping for needed tools and information. The tools can even be designed to pass values entered in one tool into other system tools and to securely save data for future sessions. Next generation systems will go much further. For example, knowledge management capability will help farmers to store and harness large amounts of information. Visualization and simulation capability should be integrated into the systems, easing their routine use.

Data needed for visualization and simulation won't have to be located; it will be automatically drawn from the knowledge base. Expert systems will help identify pest species and suggest strategies to manage them. Other systems will help manage financial risk or optimize weather predictions. Integrated farm management systems represent an advanced example of cognitive growth.

Farming decisions involve large amounts of information from different disciplines, constant uncertainty and high financial risk; systems are needed soon to help farmers.

Efforts needed by Policy Makers

For extensive use of satellite based Web extension there is a need to bring some changes in policy are identified and argued here.

1. **Collaboration between Content** - Providers Extension services around the world will better reach and help farmers if government, NGOs and Agribusiness collaborate to contribute content. All too often however, extension staff seeking needed content face recurring problems.
2. **Infrastructure for Web Delivery to Rural Areas** - Many agencies and funding entities have been slow to adopt the Web and have continued costly projects that could be better spent on Web delivery. Sustainable models are out there such as M.S. Swaminathan's Pondicherry project and N-Logue's new Chiraag model for community kiosks, in India. Broadband access is also a question; the most challenging issue that faces web-based extension is the need for affordable broadband access for the rural world.
3. **Coordination between Funding Agencies** - Agencies that fund agricultural extension and related missions must better coordinate short term efforts. The World Bank, FAO, UNESCO, USAID and private foundations like Ford and Rockefeller do great things to aid rural development but too often independently. Redundant programs are created, efficiencies are not exploited and potential synergies are lost. Examples of successful coordination are there, but many times reported collaborations do not appear substantial. Thus, proper coordination between funding agencies are needed looking to the needs of small farmers are more important than the benefits of competition.

The role of Information Technology to develop agriculture and quality of life in rural area is well established. IT can help an average Indian farmer to get relevant information regarding agro-inputs, crop production technologies, agro processing, market support, agrofinance and management of farm agri-business. The agricultural extension mechanism is becoming dependent on IT to provide appropriate and location specific technologies for the farmers to furnish timely and proficient advice to the farmers IT can be a best mean not only to develop agricultural extension but also to expand agriculture research and education system.

Government has taken up a number of satellite based "cyber extension" initiatives, across the country. The IT based agricultural information on Net is building up slowly but surely. There is still a lot of work to be done by policy makers to understand and act on this issue because the satellite-based extension through the use of Internet facility still has a long way to go to be "user friendly" for rural people. People who work in the field of satellite based extension should have proper understanding that extension through satellite based communication in rural areas is not impossible but it requires unique administrative, participatory and organizational efforts. Many organizations are supporting a number of SAUs to digitize their technical information on the Web. At the same time research and training organizations, both in public and voluntary sectors are being supported in building their capacity to digitize the rural and agricultural information and make it available on the WWW. Efforts are also being made to connect various districts of our country on the Internet so that the farmers can get all the technical information on the Web sites. If Cyber connectivity will be made available to all India Villages to take benefits of satellite based information and communication technology, the positive change in the face of rural development will be definitely possible.

Role and Potential of Information Technology in Agricultural Extension Services

The impact of reducing government expenditure on advisory services has led to the wide scale adoption of IT by AES in an effort to contain or even reduce costs. The need to commercialise AES has also resulted in the development of IT-based products in order to offer services that are attractive to the clients of AES. The need for capital expenditure initially and in the longer term needs to be based on clear business objectives and supported by robust business plans. Most AES have adopted IT in two main areas of activity.

1. To support the advisory and extension processes in working with farmers and agribusiness. This includes the development of a wide range of applications software covering such areas as business management, animal nutrition and breeding, crop husbandry, irrigation, buildings, drainage and farm waste systems design, marketing and investment appraisal. In addition, much emphasis has been placed on information collection and dissemination of time critical information such as weather, markets, pests and diseases etc. The electronic storage and retrieval of information has also been a long term goal of AES which has met with mixed success but new Internet style technologies and reducing hardware costs seem to have overcome previous obstacles in this area.
2. AES have used IT to improve the management of their operations by investing in a range of management information systems covering the areas of finance, staff, clients and work programmes. This has enabled AES to become more responsive to clients and at the same time more accountable to their founders.

The future potential for use of IT by AES is also reviewed and consideration to such areas as modelling, decision support systems, pattern recognition and satellite-based systems. The pressures on agriculture to reduce its impact on the environment, improve its record on food safety and utilise more animal-friendly production systems will push extension services into the development and adoption of IT systems to assist farmers to meet these challenges.

Management of Information by AES

In recent years the availability of Internet technologies has led to the development of Intranet services by AES. This has proved to be more reliable and cheaper than old database technologies handling information in the form of documents which can be easily downloaded and incorporated into reports and other documents which can be passed on to the clients of AES. Information provided in this way can be categorised as follows:

- Technical and farm management information for use with clients
- Time critical information to be collected, processed and rapidly disseminated to clients Role and Potential of Information Technology in Agricultural Extension Services
- Business information made available to staff and managers about the operation of AES
- Subject matter discussion groups within AES

The information that AES collect and use becomes a resource to be managed at the corporate level and the system design must be based on a clearly defined set of business and user requirements. The system will need to be able to deliver pages of text, spreadsheets, graphics and images to end users over a variety of communication means (LANs, WANs, PSTN, ISDN). In addition, some Internet services aimed at the wider agricultural public have been developed, a good example being ZADI's DAINet.

"New" Information Services for Farmers

These services aim to provide information to farmers who pay a subscription to join them and use the service. Both services use the same private network and they aim to provide an exclusive service to farmers and agribusiness. Services are focused firmly on communication and interaction, which are the key strengths of the technology. As with videotex the main focus is still mainly on market and weather information and there are extensive e-mail facilities included in the system.

Development and Use of Applications Software by AES

The development of competitive elements in the groups as farmers to improve the physical and financial performance of their enterprises (league table approach). Other applications have covered the area of design software (CAD), for farm buildings, drainage and irrigation systems, livestock handling pens, packhouses, farm waste systems, glasshouses and even to the design of golf courses. Recently there has been much interest in decision support systems (DSS) mainly in the field of arable crop production covering such areas as variety selection, fertiliser applications, pest and disease control and weed control. AES have been at the forefront of the development of these systems, which in many cases have been based on physiological models developed by researchers.

Systems Business Classification

It was important from a strategy management viewpoint to classify the systems in their contribution to the overall business aims of ADAS. The classification is as follows:

- Support systems that are valuable but not critical to current success. The major benefit is economy
- Operational systems on which ADAS currently depends and for which the main benefit is efficiency
- Business impact systems that are critical to support the present and future success of the business. The major benefit here is effectiveness
- Opportunity systems which may be important in the future. Allowance is made for innovation and technology in a forward-looking manner.

Each requires a different style of management. For example, support systems require low risk solutions with proven technologies with justification based solely on tangible return on investment with optimal use of resources. Operational systems are automating the primary business activities and the systems should deliver solutions with a good fit to the essential business requirements. Integrity and cost effectiveness are the key components. Business impact systems require users to play a central role and the emphasis should be on problem solving using new technologies where appropriate and accepting an element of risk. Opportunity systems however, are often user initiated and contain high levels of risk. Again they are driven from a problem solving viewpoint and represent essential investment in exploring forward. There needs to be an encouragement of innovation and risk taking but control by stringent reviews (Houseman 1991).

Use of IT for the Management of AES

All AES have also invested in IT for their own management purposes and have developed systems which will provide useful information to managers assisting them to improve the effectiveness of AES through the achievement of cost reductions and greater focus and clarity of purpose.

The basic requirement is to have information on 5 key areas:

- clients - farmers and other users of the extension service
- finances - budgets and expenditure
- personnel - staff information
- work - activities and programmes
- resources - offices, computers etc.

Clearly all of the above areas are linked in many different ways and are underpinned by the structure of the extension service. Any systems that are developed will need to interface with any existing Management Information System and should be simple, cheap and effective to operate. In the initial stages it likely that certain areas e.g. finance would be given higher priority than others. The systems should serve the needs of managers at all levels and produce information designed to help them manage their part of the organisation.

All extension services need to collect and keep up-to-date a minimum set of information on their clients such as name and address, telephone number, size of farm and so on. The temptation here is to collect and store a vast amount of information about each farm, its cropping and stocking, how many tractors it has and of what make, its soils, crop yields, fertiliser use etc.

The budgeting and monitoring requirement for most AES could easily be met by simple standard spreadsheets. Alternatively, PC-based accounting packages are freely available and offer value for money. Aggregation of expenditure up through the organisation should also be relatively easy to achieve. The personnel of any extension service are its most valuable asset and it is highly desirable that a database of the extension service staff is constructed. The basic personnel data requirements such as name, location, qualifications and skills should be mandatory and that this data set should be extended to cover staff appraisals and annual report information.

It is recommended that basic intermediate information about the number of farmers visited or the number of farmers attending a meeting should be collected as a routine but information about those farmers e.g. the area of their farms is only collected when a more detailed analysis of extension effectiveness is required. Systems for the future with agriculture coming under increasing pressure, not only from an economic standpoint, but also from the many public concerns such as the environment, food safety and animal welfare, the demands on AES to develop IT systems to assist farmers adjust to these pressures is bound to increase.

The main areas of current interest would appear to be as follows:

- models for the environment and ecology
- DSS for management and information intensive production
- GIS for environmental and ecological mapping and pollution
- pattern recognition for quality parameters and pest and disease identification
- satellite and GPS systems for "precision agriculture
- food product tracking through the food chain Role and Potential of Information Technology in Agricultural Extension Services
- animal tracing

■■■■

Chapter - 15

Planning and Evaluation of Extension Programmes

Extension Programmes

The all extension activity requires careful planning if it is to be effective. No extension activity is planned in isolation; every demonstration, public meeting or film show is part of an overall extension programme through which an extension agent and farmers work toward the agricultural development of their area.

An extension programme is a written statement which contains the following four elements:

1. **Objectives** - Which the agent expects to be achieved in the area within a specified period of time. This will often be a one-year period, to enable the agent to review the programme at the start of each farming year.
2. **Means** - of achieving these objectives.
3. **Resources** - that are needed to fulfil the programme.
4. **Work plan** - indicating the schedule of extension activities that will lead to the fulfilment of the programme objectives.

An extension programme with clearly defined objectives is helpful to local farmers, the agent himself, his senior extension officers and other rural development agencies. For the farmers, it shows both what they can expect from the extension service and how effective the agent is. For the agent, the programme provides a firm basis for planning extension activities on a monthly and weekly basis and for anticipating well in advance what resources will be needed. Senior extension officers can use programmes to assess agents' performance, to offer advice for improvement and to justify requests for additional staff, equipment and funds. Furthermore, the programme helps other agencies to coordinate their activities with what the agent is doing. However, programmes can only be used in these ways if they are written and made available to all concerned. A written programme is also useful when staff changes bring a new agent into the area. The new agent can use the programme to carry on from where his predecessor finished, thereby ensuring continuity of activities.

All organizations involved in agricultural development have their own procedures for planning, which can vary considerably. In particular, they can differ in the extent to which plans are made at national or local level. When considering the planning of extension programmes, two different forms can be distinguished.

Planning from below

Farmers, with their extension agents, make plans for developing local agriculture on the basis of local needs and potential and then make requests for specific assistance from national and regional authorities.

Planning from above

The agent is simply expected to implement plans made at national level. For example, He may be given a target number of hectares to be planted with improved seeds, or a specified number of farmers' groups to set up.

Successful extension programmes should include both planning approaches. National programmes will also make funds and inputs available for particular kinds of activity, which will influence the agent in his local planning decisions.

Stages in programme planning

Whatever particular procedures for programme planning are laid down by the extension organization, the following five distinct stages can be identified.

Analyse - the present situation.

Set - objectives for the extension programme.

Develop - the programme by identifying what needs to be done to achieve the objectives, and then prepare a work plan.

Implement - the programme by putting the work plan into effect.

Evaluate - the programme and its achievements as a basis for planning future programmes.

The above different stages of extension programme planning are interrelated and the planning does not always proceed neatly from one stage to another. Provisional objectives, for example, may be set during the situation analysis stage, but they may later be altered as new facts are collected and a deeper analysis leads to proper understanding of the situation. Objectives may have to be altered still further as unexpected difficulties arise while the programme is being implemented. However, it is useful for the agent to think of programme planning as involving these five activities, each of which will be considered in more detail, as each can be broken down into smaller steps.

Situation analysis

Before an extension programme can be draw up, the existing situation must first be analysed. Farming problems and their causes must be understood and the natural, human and other resources of the area identified. This stage involves three activities.

Collecting facts

The agent will need information in order to have a good understanding of the situation with which he is dealing, for example, on social structure and local culture, farming systems, education and literacy levels, size of farms, local channels of communication, transport facilities, local credit systems, marketing, health and nutrition levels, and crops and livestock.

These facts can be collected from a variety of sources. Reports of soil classification and land-use surveys, farm management studies, social surveys and previous programmes can provide a lot of useful background information. If the agent keeps good records of the farms in his area, he will have at his fingertips much of the information he needs. He can also obtain a lot of his information from farmers and local leaders. At public meetings, in group discussions and in contact with individual farmers, the agent should listen, ask questions and gradually build up a fuller understanding of the social, agricultural and economic features of the area.

Analysing facts

Facts do not speak for themselves. If farmers report that yields have declined in recent years, the agent must look for other information that would suggest an explanation. Is it because of low rainfall, declining soil fertility, or poor seed? The agent must also separate fact from opinion and guesswork. He may obtain conflicting information from two different sources and must judge which is the more reliable.

Identifying problems and potential

Farmers may know what their problems are, but the agent can bring his own perception of local problems based on a more scientific understanding of farming. He will be able to explain problems more fully and relate them to processes which farmers may not realize are in any way connected. Because of his training and experience, he will also have more suggestions to make about how the resources of the area could be used more productively.

Setting objectives

Once the existing situation has been analysed, decisions can be made about the changes that should be brought about through an extension programme. The key questions are how will local problems be solved and how will local potential be developed. Solutions will require clear, realistic objectives which should be set in following three stages.

1. Finding solutions

In looking for solutions to local problems, the agent should distinguish between technical solutions, involving improved inputs or simple changes in husbandry practice, and solutions which involve institutional changes, such as improved credit and marketing systems. Solutions involving institutional changes may require action by other agencies and at higher levels. While the agent should certainly suggest such solutions to those responsible, there may be little that can be done locally in isolation.

The sources of ideas for developing an area's potential include:

- the agent's own technical knowledge;
- farmers and agents from other areas who have tackled similar problems successfully;
- applied research which tests new ideas under farm conditions;

- national priorities and directives;
- projects which make funds available for particular activities.

2. Selecting solutions

When selecting from among the range of solutions and possible improvements, agent and farmers should ensure that proposed solutions are:

Acceptable - to farmers in the area.

Technically - sound and tested by research and experience elsewhere.

Consistent - with national policy, and with the local activities of other agencies.

Feasible - within the time and with the resources available to farmers and the extension service.

Within - the scope of the agent's ability and job description.

The agent may find that some problems will have no feasible or acceptable solution that can be implemented locally within the period of the extension programme. They may require legislation, action at other levels and by other agencies, or more research. The agent should lessen the effect of such problems where possible and act as a channel for putting forward the case for changes to those who have the power to make them.

3. Stating objectives

It should now be possible to state what the objectives of the extension programme are to be. But because his time and resources may be limited, the agent must decide which objectives have a higher priority than others. In doing so, he should consider national priorities and the size and distribution of the benefits that will arise from a given input of time and resources.

The objectives for an annual extension programme will state what should have been achieved by the end of the programme. These statements can be broken down into intermediate steps to be taken during the year in order to achieve the programme objectives. Again, the agent will have to make choices, selecting the most appropriate steps from several possibilities.

DEVELOPING AN EXTENSION PROGRAMME

Problem	**Local shortage of staple food.**
Potential	Some local farmers have increased maize yields by 30-40 percent by using improved seed and farm yard manure most farmers have plenty of manure but do not use it.
Solutions	A. Increase in size of small farms and introduce labour-saving techniques - not feasible: no access to additional land. B. Encourage larger, mechanized farms to grow more food crops - not acceptable: poorer farm families could not buy the food. C. Enable smaller farmers to increase yields - feasible, using better varieties and tested husbandry improvements.

Problem	**Local shortage of staple food.**
Preferred solution C	
Objective	Raise yields by 30 per cent on 20 per cent of the farms in the area in the first year.
Intermediate objectives	50 percent of farmers will learn of the benefits, and will acquire the skill, of using manure.
	50 percent of farmers will learn the benefits of new varieties.
	20 percent of farmers will plant improved varieties.
Plan of work	- Establish demonstration plots on ten farms.
	- Hold public meetings and film shows in ten villages to show the benefits of new varieties and improved husbandry.
	- Hold method demonstrations of manuring and correct spacing at the appropriate times.
	- Visit farms during planting season.
	- Hold result demonstrations on selected plots to encourage more farmers to try the new varieties and practices next year.
Support needed	- Subject-matter specialist to speak at public meetings, attend demonstrations and answer farmers' questions.
	- Adequate supplies of seeds, at the right time, at local stores.
	- Farm broadcasts to include relevant items at the appropriate time of year.

When the planning is completed for other programme objectives, the agent can compile all the plans into an overall annual work plan.

Implementation

To implement the programme, the agent carries out the activities specified in the work plan. His detailed monthly and weekly plans will take account of progress and problems encountered in previous months. For example, the timing of some activities may have to be changed, or additional method demonstrations may be planned if more farmers than expected want to take part. An extension programme should be flexible enough to allow the agent to respond to circumstances in this way.

Evaluation

The agent will be constantly reviewing and evaluating his progress during the year. At the end of the year, a more thorough evaluation should be carried out in which the agent identifies how fully each objective has been achieved, and the reasons for any lack of progress. This evaluation, together with an up-dated situation analysis, provides the basis for planning the next year's programme.

Evaluating Extension Programmes

Evaluation is the process by which the effectiveness of extension is assessed. It is more than simply finding out what happened; it involves passing judgement on what happened. Was the outcome of the programme good enough? Was it better or worse than expected? Could more have been achieved?

Extension programmes are evaluated to

a. ascertain for the extension organization how well agents perform, so that their suitability for promotion may be assessed.

b. satisfy the government that public money spent on extension is being used effectively.

c. permit the agent to learn from what has happened. Evaluation is a waste of time unless the results have an influence on future extension decisions.

Evaluation is an important first step and one which is made easier if the programme had clear, precise objectives. If the answer is no, then there is no real basis on which to make improvements in future programmes. It is important, therefore, to ask why things turned out in the way they did. Only when that question is answered can the agent learn from the past. Agents should, therefore, ask questions about the following aspects of the programme.

Results. What happened as a result of the extension programme? Were they the results that were expected, and were there any unexpected results?

Inputs. Were all the planned inputs available and, if not, why?

Levels of Evaluation

There are several levels of evaluation of extension programmes. At the most general level, the effect of extension on agricultural production, family incomes and standards of living can be evaluated. An increase in family living standards is usually an important ultimate goal of rural extension. Evaluation of this kind involves measuring production and farm income for a representative sample of farm families, and then comparing the figures with previous levels.

An intermediate level of evaluation is provided by the extension programme itself. Two questions are important here. Did the extension activities take place in the planned sequence and at the right time? Did these activities lead to the expected results? If the answers are negative, the agent should try to understand why. Finally, the agent can evaluate at the level of each extension activity. All extension activities, such as demonstrations, talks or meetings with a farmers' group, have a purpose. The agent should try to check, wherever possible, not only how well the activity itself was conducted but whether the purpose was achieved. This will usually involve finding out whether the extension activity led to any changes in one or more of the following:

– awareness of a particular idea, possibility or problem;
– motivation to act in a particular way;
– knowledge about new farming practices;
– skills needed to adopt a new practice;
– behaviour by farmers and their families (such as new farming methods), or by an extension group (such as making an application for funds to carry out a group project or the preparation of a formal group constitution).

At all levels of evaluation, the agent needs to collect information to compare the situation after the activity with the situation existing before. When planning a result demonstration, for example, he will have some idea of how many farmers in the area know about, are interested in, or have already adopted the particular practice that is to be demonstrated.

There are several ways of collecting information for evaluation at the village level.

Agent's reports: Each extension activity is required from agents by their extension officers, the agent should make some notes on each activity for his own use, concentrating on his conduct of the activity and on points to note for future occasions.

Supervisors: It is not easy for an agent to assess the extension activity. It is useful, therefore, to have constructive comments from a supervisor or colleague.

Discussions: Informal discussion with farmers after the extension activity will reveal their immediate reactions.

Questionnaires: Simple check-lists and questionnaires can be used when the agent has the time and opportunity to carry out a more formal evaluation of extension activities.

Observation: Where changes in farming practices, observation is an accurate source of information.

Many extension organizations have their own formal procedures for evaluation. The agent prepares a detailed plan of work each month on a standard form. This procedure may be combined with a monthly meeting of agents in a particular district at which progress and problems in each area are discussed.

Key Performance Indicator (KPI)

PKI define factors the institution needs to benchmark and monitor. Assessment techniques provide the mechanism for measuring and evaluating the defined factors to evaluate progress or impact. PKIs specify what is measured and assessment techniques detail how and when it will be measured. KPI is a measure used to define and evaluate how successful an organization is. Typically is expressed in terms of making progress towards its long-term organizational goals. KPI incorporates information on the sources, calculations and definitions for each measure and sets out the timetable for submission of monthly data.

PKIs assist an organisation to define and measure progress towards organisational goals and objectives. Once an organisation has analysed its mission and defined its goals, it needs to measure progress towards those goals. KPIs provide a measurement tool.

A performance indicator or key performance indicator (KPI) is a type of performance measurement. KPI evaluate the success of an organisation or of a particular activity (such as projects, programs, products and other initiatives) in which it engages. These assessments often lead to the identification of potential improvements so performance indicators are routinely associated with performance improvement initiatives. A very common way to choose KPIs is to apply a management framework such as the balanced scorecard.

■■■■

Chapter - 16 Agricultural and Rural Development Programmes in India

In India more than 80 per cent peoples are living below poverty line. Therefore, several experiment was undertaken by government and non-government agencies which proved the beneficial efforts of intensive area projects in which activities in various field of development were closely integrated. During the first plan several rural development plans are introduced aimed at initiating and directing a process of change with a view to transform the social and economic life of villages.

Historical Development of Extension and Rural Development Programme in India

The term extension was first used in the United States of America in 1862 by J. Paul Leagans.

Historical development can be described in 4 stages:

1. Pre independence era (1866-1947)
2. Post independence era (1948-1952)
3. Community development and national extension service (1953-1960)
4. Intensive agriculture development era (1961 - onwards)

1. Pre Independence Era

During the pre independence period various scattered and short lived efforts were made towards rural development in various parts of the country by individuals or some organizations.

(i) **Shriniketan:** This program was started in Bengal in 1908 by Ravindra Nath Tagore. Tagore believed that there was need for cultural development. In view of this programme was launched. This programme was known as 'light through learning'.

The following were the objectives of this programme:

- To develop interest among rural people through rural developmental work.
- To analyse the rural people so that problem and execution of possible situation.
- To develop national resources of rural people so that they can do their work independently.
- To develop cooperativeness among rural peoples.

(ii) Gurgaon project: This project was started by F.L. Brayne in 1920 at Punjab. The aim of this project was to make the people aware that development is possible by changing their behaviour with the adoption of new agricultural practices, improving their health standard, woman education and home development work.

(iii) Sevagram project: This project was started by Gandhiji in1926 at Gujarat. The main principle was truth and non violence for social and economical development. The other objectives were Hindu Muslim unity and basic education through learning by doing.

(iv) Marthandam project: This project was started by Dr. Spensar Hatch in 1928 at Kerala. This was also known as Young Men Christian Association (YMCA) movement. The objectives were set to develop self help of the people with guidance and strength through fellowship.

(v) Grow More Food Campaign: This was started by Govt. of India in 1942 in Bengal.

(vi) Indian Village Scheme: This was started by W.H. Wisher in 1945. The objectives were

- Realization of abundant life and to make villagers able to solve their own problems.
- To motivate people regarding cooperation for development work.
- To give a chance to rural people for their help in rural upliftment work.

(vi) Firka Development Scheme (FDS): Mr. Raj Gopalachari's idea was taken by Govt. of Madras in 1946-47.

Post independent era: Following programmes were started:

1. Etawah Pilot Project started in 1948 by Albert Mayer
ii. Sarvodaya Program: Achrya Vinoba Bhave started this programme in 1948 in Bombay.
iii. Community Development Programme started by Govt. of India in 1952.

Period of 1953 – 1960: Number of programmes were launched during this period. These are:

i. National Extension Service started by Govt. of India in 1953
ii. Panchayati Raj
iii. Intensive Agricultural District Programme

Period of 1961 – onwards

i. High Yielding Variety Programme
ii. National Demonstration Project: Initially this programme was started in 100 districts of the country in 1965.
iii. Small Farmers Development Agency
iv. Marginal Farmer and Agriculture Labour
v. Drought Prone Area Programme
vi. Training and Visit.

Important systems of extension in each of the stages, with their commonly used abbreviations and year of initiation are presented in next page:

Community Development

1952	CDP	Community Development Programme
1953	NES	National Extension Service
1954	CDB	Community Development Block
1957	Panchayati Raj	Democratic Decentralization

Technological Development

1960	IADP	Intensive Agricultural District Programme
1964	IAAP	Intensive Agricultural Area Programme
1964-65	ICDP	Intensive Cattle Development Project
1966	HYVP	High Yielding Variety Programme

Development with special justice

1967 (26 Jan)		Krishi Darshan
1970-71	SFDA	Small Farmers Development Agency
	MFAL	Marginal Farmer's and Agricultural Labourers Project
	DPAP	Drought Prone Area Programme
1971-72	TADP	Tribal Area Development Project
1972-73	PPTD	Pilot Project for Tribal Development
1974	T&V	Training and Visit
1974	MFAS	Mobile Farm Advisory Service
1975	FS	Farm School
1977	AY	Antyodaya Yojna/Movement
1978-79	IRDP	Integrated Rural Development Programme
1979	ORP	Operational Research Projects
1979	LLP	Lab to Land Programme
1979	TRYSEM	Training of Rural Youth for Self – Employment
1980	NREP	National Rural Employment Programme
1982	DWCRA	Development of Woman and Children in Rural Areas
1983	NAEP	National Agricultural Extension Project
1985	CIS	Crop Insurance Scheme
1986	TMO	Technology Mission on Oilseeds
1989	JRY	Jawahar Rozgar Yojna

1993	EAS	Employment Assurance Scheme
1994	SFAC	Small Farmers Agri-business Consortium
2005	ATMA	Agriculture Technology Management Agency
2007	NFSM	National Food Security Mission
2015	PMKSY	Pradhan Mantri Kisan Sinchai Yojna
2016	PMFBY	Pradhan Mantri Fasal Beema Yojna
2019	PMKSN	Pradhan Mantri Kisan Samman Nidhi

S.N.	Name of project	Founder	Location	Year
1.	Co-operative movement	Nicholson		1904
2.	Sriniketan	R.N. Tagore	West Bengal	1908
3.	Gurgaon	B.L. Brayne	Punjab	1920
4.	Sevagram	M.K. Gandhi	Gujarat	1926
5.	Marthandam (YMCA)	Spensor Hatch	Kerala	1928
6.	Baroda project		Gujarat	1932
7.	Social Legislation	Govt. of India		1929
8.	Rural Development in U.P.	State Govt.	UP	1937
9.	Grow More Food Campaign	Govt. of India	Bengal	1942
10.	Indian Village Service	W.H. Wisher	All village	1945
11.	Firka Development Scheme	Govt. of Madras	Madras	1946-47
12.	Nilokheri	Govt. of India	Haryana	1948
13.	Etawah Pilot Project	Albert Mayer	UP	1948
14.	Sarvodaya Programme	Vinoba Bhave	Bombay	1948
15.	Community Development Programme	Govt. of India	India	1952
16.	Panchayati Raj	Govt. of India	India	1957
17.	Intensive Agriculture District Programme	Govt. of India	India	1960-61
18.	High Yielding Variety Programme	Govt. of India	India	1966
19.	National Demonstration Project	Govt. of India	India	1965
20.	SFDA & SMFAL	Govt. of India	India	1970-71
21.	Drought Prone Area Programme	Govt. of India	India	1970-71
22.	Training & Visit	Govt. of India	India	1974
23.	Agriculture Technology Management Agency	Govt. of India	India	2005
24.	National Food Security Mission	Govt. of India	India	2007

Integrated Rural Development Programme (IRDP)

- The strategy for rural development adopted in the sixth plan is known as the Integrated Rural Development.
- The integrated rural development focused on the target group comprising small and marginal farmers, agricultural labourers and rural artisans, where economic improvement was on important concern of rural development.
- The IRDP aimed at integrated field programmes reflecting the economic activities of the rural families, whose employment and development were the basic objectives.
- It was proposed to bring this about by developing the primary, secondary and the tertiary sectors.
- In the primary sectors, programmes for agriculture, animal husbandry, fisheries and forestry development were intensified.
- In the secondary sector, programmes for village and cottage industries as well as small scale industries and supporting services were substantially enlarged and strengthened.
- The tertiary sector was developed by creating facilities for organized marketing, processing and allied activities so as to absorb an increasing number of local people.
- Programme of agriculture, animal husbandry, fisheries and forestry development were brought into focus by maximizing land and water use and cropping facilities in a holding through a mixed farming approach.

This was the core of the I.R.D. Programme for implemented at the district level. District Rural Development Agencies have been set up an all the district in the country. Each block identifies some 600 families which have been benefited in the year 1981-82 under the programme. IRDP was launched in 1978-79.

Intensive Agriculture Area Programme (I.A.A.P.)

The Intensive Agricultural Area Programme was initiated early in 1964-65 for an intensive development. The objective of the Intensive Agriculture Area Programme was to increase production of the main crops in selected areas by an intensive and coordinated use of various aids. By the observation of the mid term appraisal of the third plan much emphasis shall be given to the development of scientific and progressive agriculture in intensive manner is an areas where a high agriculture production potential exists. In the third plan, the Agriculture Production Board in the Union Ministry of Food and Agriculture had agreed that about 20 to 25 per cent of the cultivated area of the country should be selected for an Intensive Agriculture Development.

The IAAP followed the package approach for the use in agriculture of inter related factors as social, physical, institutional in a strategic combination which were like to exert an impact of agricultural production. The difference between the I.A.A.P. and I.A.D.P. (Intensive District Agriculture Programme) was that in the I.A.A.P. there was a lower degree of intensity in respect of the staffing pattern, supervision, establishment of soil testing laboratories and seed testing centres. Ultimately the I.A.A.P. has also been developed to the same level of intensity as that of I.A.D.P.

Community Development Programme (CDP)

Community Development Programme has been defined as "a movement designed to promote better living for the whole community with the active participation and on the initiative of the community".

The change in rural life has been thought to be brought about by concerned efforts and an integrated approach making an assault on all facets of rural life through a single multipurpose agency *viz.*, C.D. Projects. The project widely recognizes as:

a. Triple Tenets of Faith

1. Muscles can do it
2. Muscle can be trained to do it
3. Conditions can be created to do it

b. Triple Character of Rights

1. Right to lie
2. Right to learn
3. Right to receive what is earned

c. Triple Rules of Business

1. Elevation
2. Integration
3. Standerdization

The Community Development Programme was launched on 2nd October, 1952 with 55 PILOT projects on an experimental basis. The programme was implemented in units, each block covers an area of about 400 to 500 sq. km, with about 100 villages and a population of about 1 lakh. There were 5026 community Development Blocks, covering a population of about 47 crores spread over all the 6 lakh villages in the country. The finances for the programme were drawn both from the people and the government. The C.D. Programme being essentially a people's programme, prescribed a qualifying scale of voluntary contribution from the people in the form of labour as well as money. The balance expenditure in a block was met by the state, the expenses were shared by the Central Government and the State Governments. State Governments were free to provide resources for the programme on priority basis. The Central Government's assistance to the states was given in the form either of loans or of annual lumpsum grants. C.D. Programme awakened consciousness in the people and made than articulate for needs which they had not felt before.

Marginal Farmers and Agricultural Labours Agency (M.F.A.L.A.)

Marginal Farmers and Agricultural Labours Agency were set up on the recommendation of the Rural Credit Review Committee (1969). The principle objective of the M.F.A.L.A. programme was to assist the marginal cultivators in making the maximum productive use of their small holdings by undertaking animal husbandry, dairy and horticulture etc. The efforts were directed towards generating labourers income by channelising credit, improved inputs and improved practices. It assisted other institutions, primarily cooperative societies, to implement these programme.

The main functions of M.F.A. L. Agency were:

1. To identify marginal farmers and agricultural labourers to be covered by the project.
2. To investigate their problems.
3. To facilitate economic programmes for providing gainful employment to participants.
4. To create facilities for production and marketing of agricultural products.
5. To arrange institutional financial administrative help for adoption of various programme.

Small Farmers Development Agency (SFDA)

Small Farmers Development Agency was established to help small farmers in selected areas. Small farmers Development Agency were firstly established in 46 selected district on a pilot basis. The main objectives of this project was to help the small farmers to become viable over a period of 3 to 4 years. The project was mainly helping in production and credit facilities. Each agency was meant to benefit about 50000 potentially viable farmers through assistance in the form of subsidized irrigation, soil conservation, drainage, processing and marketing facilities.

The main functions of S.F.D.A. were as follows:

1. To identify small farmers.
2. To identify problems and ways of implementing these programme.
3. To supply various inputs required by the small farmers.
4. To help in credit facilities.
5. To help farmers for transportation, storage, processing and marketing of their produce.
6. To arranged irrigation facilities for small farmers.

A sum of Rs 210 lakhs was allotted for the projects in the fifth Five Year Plan, during the year 1977-78 a sum of Rs 70 lakhs has been provided in the budget. Encouraging results have been achieved in this project with regards to agricultural demonstration, multiple cropping and introduction of high yielding varieties.

Drought Prone Area Programme (DPAP)

The Drought Prone Area Programme was initiated during the fourth Five Year Plan in 1970-71. The drought prone areas were selected on the basis of unirrigated areas, low or erratic distribution of rainfall and high periodicity of drought.

The basic objectives of the programme were:

1. To minimize the impact of severity of drought.
2. To utilize all resources in the affected areas.

3. To improve the standard of living of rural drought affected people.

The main plank was to restore a proper ecological balance in drought prone areas through:

1. Development and management of irrigation resources.
2. Soil and water conservation and afforestation programme.
3. Re-structuring of cropping pattern and pasture development.
4. Livestock development.
5. Development of small and marginal farmers.

The drought prone areas represently occupied nearly 19 per cent of the total area of the country and account for nearly 12 per cent of the population. During the fifth plant period, Rs 181.50 crores had been allotted for DPAP of the minor irrigation scheme would account for 60 per cent, rural communication for about 25 per cent and remaining 15 per cent for soil conservation, afforestation and drinking water supply.

Minimum Need Programme (MNP)

This programme redefined the norms of nine items of minimum needs which were determined for the fifth plan. The items included are:

i. Elementary education for children upto the age of 14 at the nearest possible place to their home.

ii. Adult education

iii. Rural health

iv. Rural water supply

v. Provision of roads to all villages having population of 1500 persons or more, this minimum limits being considered for a cluster of villages in case of hilly tribal and coastal areas.

vi. Rural electrification

vii. Provision of developed home sites for landless labourers in rural areas.

viii. Environmental improvement of urban slums

ix. Nutrition programme.

The above norms were subsequently upgraded and the programme was renamed as: "Revised Minimum Need Programme" (RMNP). In RMNP, the elementary education, adult education and nutrition programme relate to urban as well as rural areas. The programme for the environmental improvement of slums related to urban areas. All other programme were designed to cater only to the rural needs. Urban needs in these sectors were proposed to be covered under general secretarial plans.

Tribal Area Development Project (TADP)

Tribal Area Development Project initiated in 1971-72 for the development of six selected tribal areas in Andhra Pradesh (Srikakulum), Bihar (Singh Bhum), MP (Dantewada, Korba) and Orissa

(Ganjam, Koraput). Under this programme, special Tribal Development Agency (TDA) were given chance to study the problems of tribal areas relating to the economic development, communication and administration, agriculture and allied activities.

Two more project in Orissa were taken up as part of the fifth plan, Rs 12 crores had been provided for these 6 TDA projects. By 1977-78 the total expenditure was Rs 12.05 crores since inception. The projects utilized Rs 10.41 crores upto 1977-78. These projects have so far identified about 353384 tribal participants of which 293012 were benefited under the different economic development programme. About 5.8 lakh hectares land has been brought under improved agriculture use for cultivation of various crops.

Antyodaya Yojna or movement

Antyodaya means that upliftment of the last man in the row. The major aims at economic upliftment of the poorest of the poor families in every village of Rajasthan. The following distinguish feature of this project are:

i. Criteria of selection is poverty irrespective of caste, religion, domicile etc.

ii. Complete absence of regional disparity as its impact is to be felt by all the villages.

iii. The coverage starts from bottom and travels upward.

iv. There qualitative change in the basic approach of implementation. Government machinery and financial institutions approach the identified families right at their doorsteps.

v. The programme covers a wide range of scheme e.g. pension for disabled, land (if available) for landless, jobs etc.

vi. The focus is an individual family.

The programmes was launched on 2 October, 1977. In the first stage, five families per village were selected. In the second stage, which commenced on October 2, 1978 a graduated scale for identification of number of Antyodaya families were used. The number of families to be selected under the scheme from each village varies with the population of that village.

It was

- 3 families from each village less than 500 population
- 5 families from each village more than 500 upto 1000 population
- 10 families from each village more than 1000 upto 1999 population
- 10 from other villages

Proper identification of the poorest families in each village was considered the crucial demand in this whole scheme. After finalizing the names of the families, the needs of these families were decided. Each identified Antyodaya family was provided with an identity card so that they may be given priority in employment jobs, rural works etc.

In the first round 1.52 lakh families were identified, out of which 1.42 lakh families were benefited upto March 31, 1980. In the second round 1.09 lakh families were identified out of which 0.86 lakh families were benefited March, 1981. In addition to subsidies made available to these families a sum of Rs 176.31 million provided as loan to them through cooperative banks, commercial banks, khadi board and industrial departments.

Food for work Programme

The basic objectives of the programme were to generate additional employment in rural areas through utilization of available stock of food grains and to create there durable community assets which could strengthen the infrastructure in rural areas for socio-economic development. Besides being successful in achieving is objectives, the programme resulted in many other benefits to the rural poor like price stabilization of food grains, ensuring minimum wages to the workers and providing work to needy poor throughout the year, creating durable community assets in the rural areas resulting in improvement in the quality of life of the rural people and arresting the exodus of rural population to cities and towns.

National Scheme of Training of Rural Youth for Self-Employment (TRYSEM)

TRYSEM was initiated from August, 1979 with the principal objectives of removal of unemployment among youth. The main thrust of the scheme is to equip rural youth with necessary skills and technology to enable them to take up self employment. The strategy for training includes, apart from institutional training, training through local servicing and industrial units, master craftmen, artisans and skilled workers. During the training period the trainees are also given financial assistance. They are also helped to prepare project reports which are converted into bankable scheme. They are helped to apply for basic loan and subsidy.

Jawahar Rozgar Yojna (JRY)

The JRY announced on April 23, 1989 will be implemented and administered by village panchayat all over the country. About 440 lakh families which are below the poverty line shall be benefited from the scheme. Existing wage employment programme *viz.*, NREP and RLGEP have been merged into this yojna. This scheme expected to give employment to atleast one member of each poor rural family for 50 to 100 days in a year. 15 per cent of total resources transferred to village panchayats will be exclusively used for the benefit of the scheduled caste and scheduled tribes. 30 per cent of the beneficiaries will be women. 80 per cent of total finances will be defrayed by the central government and remaining 20 per cent will be borne by the state government. The funds will be allotted to state and union territories in proportion to the number of persons living below the poverty line. It has been decided to give specific consideration of hills, deserts and island areas. The criteria for the distribution of resources to village panchayat from the districts on the basis of population of each village panchayat. It is expected that the village panchayat with a population of 3000 to 4000 will receive the money between Rs 80000 to 400000. 30 per cent of the employment opportunities under the yojna would be reared for women.

The primary objectives of the programme was generation of additional gainful employment for the unemployed and under employed persons both men and women in the rural areas. The secondary objectives were:

i. Creation of productive community assets for direct and continuing benefits to the poverty groups and for strengthening rural, economic and social infrastructure, which would lead to rapid growth of rural economy and steady rise in the income levels of the rural poor.

ii. Improvement in the over all quality of life in the rural areas. The special feature was that this programme was implemented all over the country through the village panchayat.

Employment Assurance Scheme (EAS)

The Employment Assurance Scheme was launched in 1993. The scheme envisaged that those who were in need and sought employment would get assured wage employment for 100 days during the lean agricultural season. The primary objective of EAS was to provide gainful employment during the lean agricultural season in manual work. The secondary objectives were the creation of economic infrastructure and community assets for sustained employment and development. The expenditure under the scheme was shared between the centre and the state on 80:20 basis. The central assistance was released directly to the District Rural Development Agencies (DRDAs).

The scheme was implemented in identified backward blocks of districts situated in the drought prone areas, desert areas, tribal areas and hilly areas in which the Revamped Public Distribution System (RPDS) was in operation.

Small Farmers Agri-business Consortium (SFAC)

SFAC was initiated in 1994. Increased production and productivity, value addition and efficient linkages between producers and consumers were high priority goals of the consortium. SFAC would undertake to fund research and development activities through the Council of Scientific and Industrial Research (CSIR) and other scientific establishments in India to promote its activities. SFAC will deal with agriculture including animal husbandry, fisheries and sericulture, apart from crop husbandry and horticulture.

Eleven districts have initially been selected for preparation of district plans under the project namely: Dharampuri (TN), Pondicherry, Ernakulam (Kerala), Bikaner (Rajasthan), Bubar (Gujarat), Midnapur (WB), Dharwar (KN), Pithoragarh (Uttarakhand), Ganjam (Orissa), Barapet (Assam) and Jabalpur (MP).

Different institutions are providing credit support in a segmented manner. Agricultural inputs, food processing, agro-forestry, agro-industries, agro-exports etc are getting finance through different institutions. SFAC will help build-up organisations and institutions which could organize farmers so that they would benefit from various schemes of agri-business and help farmers to market their product.

High Yielding Variety Programme (HYVP)

HYVP launched in 1966, which helped the country in attaining self sufficiency in food. The technological development did not remain confined to introduction of high yielding crop varieties alone. These were combined with the application of high analysis and balanced fertilizer, irrigation, plant protection, improved implements etc, which made a 'green revolution' possible in the country. High yielding technologies by themselves tend to be scale natural *i.e.* all farmers irrespective of the size of their holding can derive economic benefit from them provided they have access to the needed inputs.

Key Village Scheme

The initiation of the key village scheme in August, 1952, was the first systematic attempt to improve the quality and productivity of cattle and buffaloes in the country.

Intensive Cattle Development Project (ICDP)

The ICDP was initiated in 1964-65. The objectives of the project were the same as the key village scheme but much wider in scope. In envisaged provision of a package of improved practices to the cattle owners effect a breakthrough in milk production. The package included as cattle breeding, castration of bulls, veterinary aid and disease control, registration and milk recording, introduction of high yielding milch cattle, subsidies and incentives for milk production, feed and fodder development – demonstration, distribution of fodder seeds, silage making and popularization of chaff cutter and dairy extension.

Development of Women and Children in Rural Areas (DWCRA)

DWCRA was started as a sub scheme of IRDP in 1982-83. The objective of the programme was that women's income is known to have possible correlation with the nutritional and educational status of the family and in the building up of a possible attitude towards status of women. The objective of DWCRA is to provide income generating activities to women and also to provide an organizational support in terms of a receiving system for the assisted women so that they could become effective recipients of goods and services available in that area. The strategy was that while the target group of DWCRA was the same as IRDP, the methodology was through a group. TRYSEM is expected to fulfil the skill requirements for beneficiaries of DWCRA also.

National Agricultural Extension Project (NAEP)

The NAEP was launched in 1983. The basic objective of NAEP is to bridge the gap between well developed research system with that of the extension system, so that the transfer of technology take place at a much faster rate, resulting in higher production and prosperity of the rural sector in general and agricultural sector in particular.

Under NAEP, guidelines were developed for preliminary assessments of extension-need of each state, which were as follows:

1. In depth review of the existing agricultural extension system.
2. Statement on existing resources of Department of Agriculture and other departments that may be redeployed to further strengthen agricultural extension.
3. Review on links with supporting service, input, credit etc. and of coordination between the extension service with Department of Agricultural and related departments.
4. Review of extension/research linkages.

5. Review of adoptive research.
6. Review of training for extension agents, institutions involved, type of training and duration of course organized and future need.
7. Review of physical facilities provided in the form of accommodation, incentives to extension agents, type of mobility available to senior and field level extension functionaries.
8. Review on the manner in which an agency maintains basic crop estimation, production etc. and conduct crop cutting experiment.

Technology Mission of Oilseeds

A major deficient area was oilseeds production for which the Technology Mission on Oilseeds was initiated in May 1986. The main purpose was to make the country self-sufficient in edible and non-edible oil and to minimize their imports. This was to be achieved through an integrated approach involving different development, scientific, banking, marketing and processing agencies. The technology mission on oilseeds had the following strategy:

i. improvements in crop technology.
ii. support to the farmers to apply improved production technology.
iii. improved post-harvest technology.
iv. support to industry for applying post-harvest technology.

The comprehensive plan was prepared to increase the irrigated area under oilseeds by replacing less remunerative crops and by adopting improved production technologies and better dry farming methods and practices. The extension methods include block demonstration, training of farmers, subject matter specialist, assured input supply, technology package, field days, farmers conventions and communication through radio and TV talks.

Realizing the importance of a strong support system to farmers through transfer to technology, the Govt. of India initiated National Oilseeds Development Project (NODP) for sustained improvement in vegetable oil production.

Lab to Land Programme (LLP)

LLP was launched by the ICAR in 1979 on the occasion of Golden Jubilee celebration. The overall objective of the programme was to improve the economic condition of the small and marginal farmers and landless agricultural laboureres, particularly scheduled caste and schedules tribes, through transfer of improved technology developed by the agricultural universities, research institutions etc.

The specific objectives the lab to land programme were:

i. Study and understand the background and resources of the selected farmers and landless agricultural labourers.
ii. Assist the farmers to develop feasible farm plans keeping in view the availability of technologies, needs and resources of the farmers.

iii. Guide and help the farmers in adopting improved technologies as per their farm plans and demonstrate to them the economic viability of those technologies.

iv. Organize training programmes and other extension activities, in relation to their adopted practices and prepare them for achieve participation in agricultural development programme of the state.

v. Make the farmers aware of the various opportunities and agencies which they could utilize to their economic advantage.

vi. Develop functional relations and linkages with the scientific and institutions for future guidance, advisory services and help.

vii. Utilize this project as a feedback mechanism for the agricultural scientists and extension functionaries.

The programme was initiated with 75000 farm families over the whole country. It was implemented by the agricultural universities, ICAR institute, state department of agriculture and some selected voluntary organizations. Major thrust in the programme was to introduce the most appropriate technologies that would help in the diversification of labour use and introduction of supplementary sources of income.

Crop Insurance Scheme (CIS)

The Crop Insurance Scheme was introduced in 1985. Crop insurance scheme risk involved the substitution of a small known cost for the possibility of a large but uncertain loss. The insurance companies are able to accept the risk of loss from natural hazards on individual farms, because they are able to spread this risk over a large number of forms. As a result of spreading risk over a large number of farms, insurance companies are able to absorb the risk of loss on individual farms.

During the three years ending 1987-88, the premium collected was about Rs 42 crores against the insurance claims amounting to Rs 530 crores. According to this insurance principles, the average annual claims have to be less than or close to the annual premium.

There are following benefit by crop insurance scheme to the farmers:

1. There is big protection of farmers against the failure of crops and stabilize their income.
2. Crop insurance improves their credit worthiness in securing loans from a credit agency.
3. Farmers and trade centres improve their income stability as a result of the stabilization of farm income.
4. It reduces the government obligation of provide relief in case of crop disaster.
5. Crop insurance provide greater confidence to farmers to venture upon the adoption of modern technology involving larger expenditure on modern inputs and greater risk due to natural hazards.

Pradhan Mantra Kisan Samman Nidhi (PM - KISAN)

The Government of India launched a new Central Sector Scheme, "Pradhan Mantri Kisan Samman Nidhi (PM-KISAN)" in order to augment the income of the Small and Marginal Farmers

(SMFs). The scheme was launched in February 2019. It is being implemented by the Department of Agriculture, Co-operation and Farmers Welfare (DAC&FW) under the Ministry of Agriculture and Farmers Welfare through the Department of Agriculture of all the States and Union Territories Governments. Under the Scheme, a direct payment of Rs. 6000 per year will be transferred in three equal instalments of Rs. 2000 each every four months into the bank accounts of eligible landholding families. PM-KISAN is a central sector scheme with 100% funding from Govt. of India. At present this scheme is open for all categories of farmers.

Many measures are being taken by the Government to reach to all eligible beneficiaries under PM-KISAN. Public interfaces have been made available for self registration, checking the payment status, correction of the name as per Aadhaar as it is being mandatory requirement of the scheme. To broaden the reach further, the PM-KISAN mobile app designed and developed by National Informatics Centre (NIC), Ministry of Electronics and Information Technology, Government of India is being launched.

Using Mobile app, farmers can

- Register themselves
- Know the status about their registration and payments
- Correct name as per Aadhaar
- Know about the scheme.
- Dial Helpline Numbers

Pradhan Mantri Man Dhan Yojna

Pradhan Mantri Kisan Maan-Dhan Yojana (PM-KMY) is an old age pension scheme for all land holding Small and Marginal Farmers (SMFs) in the country. It is a voluntary and contributory pension scheme for the entry age group of 18 to 40 years. The Scheme is effective from the 9thAugust, 2019.

Salient features

1. It is voluntary and contributory for farmers in the entry age group of 18 to 40 years and a monthly pension of Rs. 3000/- will be provided to them on attaining the age of 60 years.
2. The farmers will have to make a monthly contribution of Rs. 55 to 200, depending on their age of entry, in the Pension Fund till they reach the retirement date *i.e.* the age of 60 years. The farmers will have to make a monthly contribution of Rs. 55 to 200, depending on their age of entry, in the Pension Fund till they reach the retirement date *i.e.* the age of 60 years.
3. The monthly contributions will fall due on the same day every month as enrolment date. The beneficiaries may also chose an option to pay their contributions on quarterly, 4-monthly or half-yearly basis. Such contributions will fall due on the same day of such period as the date of enrollment

4. The spouse is also eligible to get a separate pension of Rs.3000/- upon making separate contributions to the Fund.
5. The Life Insurance Corporation of India (LIC) shall be the Pension Fund Manager and responsible for pension pay out.
6. In case of death of the farmer before retirement date, the spouse may continue in the scheme by paying the remaining contributions till the remaining age of the deceased farmer. If the spouse does not wish to continue, the total contribution made by the farmer along with interest will be paid to the spouse. If there is no spouse, then total contribution along with interest will be paid to the nominee.
7. If the farmer dies after the retirement date, the spouse will receive 50% of the pension as Family Pension. After the death of both the farmer and the spouse, the accumulated corpus shall be credited back to the Pension Fund.
8. The beneficiaries may opt voluntarily to exit the Scheme after a minimum period of 5 years of regular contributions. On exit, their entire contribution shall be returned by LIC with an interest equivalent to prevailing saving bank rates.
9. The farmers, who are also beneficiaries of PM-Kisan scheme, will have the option to allow their contribution debited from the benefit of that scheme directly.
10. In case of default in making regular contributions, the beneficiaries are allowed to regularize the contributions by paying the outstanding dues along with prescribed interest. Until 1 month from first unpaid contribution, no late fee would be charged. Three payment cycles demand would be raised for payment of contribution without any interest.

Eligibility

1. Small and Marginal Farmer (SMF) - a farmer who owns cultivable land upto 2 hectare as per land records of the concerned State/UT.
2. Age of 18- 40 years.

Farmers who are not eligible for the scheme

The following categories of farmers have been brought under the exclusion criteria:

- SMFs covered under any other statuary social security schemes such as National Pension Scheme (NPS), Employees' State Insurance Corporation scheme, Employees' Fund Organization Scheme etc.
- Farmers who have opted for Pradhan Mantri Shram Yogi Maan Dhan Yojana (PM-SYM) administered by the Ministry of Labour & Employment.
- Farmers who have opted for Pradhan Mantri Laghu Vyapari Maan-dhan Yojana (PM-LVM) administered by the Ministry of Labour & Employment.
- Further, the following categories of beneficiaries of higher economic status shall not be eligible for benefits under the scheme:
    - All Institutional Land holders; and

- o Former and present holders of constitutional posts.
- o Former and present Ministers/ State Ministers and former/present Members of Lok Sabha/ Rajya Sabha/ State Legislative Assemblies/State Legislative Councils, former and present Mayors of Municipal Corporations, former and present Chairpersons of District Panchayats.
- o All serving or retired officers and employees of Central/State Government Ministries/Offices/Departments and their field units, Central or State PSEs and Attached offices/Autonomous Institutions under Government as well as regular employees of the Local Bodies (Excluding Multi Tasking Staff / Class IV/Group D employees).
- o All Persons who paid Income Tax in last assessment year.
- o Professionals like Doctors, Engineers, Lawyers, Chartered Accountants, and Architects registered with Professional bodies and carrying out profession by undertaking practice.

Deen Dayal Upadhyaya Grameen Kaushalya Yojana

Deen Dayal Upadhyaya Grameen Kaushalya Yojana, a part of National Livelihood Mission, has the objectives of catering to the career aspirations of the rural youth and adding diversity to the income of rural families. Launched on 25th September 2014, the scheme's prime focus is on the rural youth of poor families aged between 15 and 35. An amount of Rs 1500 crores has been provided for the scheme which will help in enhancing employability. The yojana is present in 21 States and Union Territories across 568 districts and 6215 blocks changing the lives of youth. Around 690 projects are being implemented by 300 partners. As per the government reports, over 2.7 lakh candidates have been trained till now and nearly 1.34 lakh candidates have been placed in jobs.

Swarnjayanti Gram Swarozgar Yojana (SGSY)/ National Rural Livelihood Mission

Swarnjayanti Gram Swarozgar Yojana which is redesigned as National Rural Livelihood Mission was launched in 2011. Also known as Ajeevika, this scheme aims at empowering women self-help model across the country. Under this scheme, the government provides a loan of 3 lakh rupees at an interest rate of 7% which can be reduced to 4% at the time of repayment. The scheme was aided by World Bank and aimed at creating efficient and also effective institutional platforms for poor people. It also helped in increasing the household income by improving access to financial services. NRLM also helps in harnessing the capabilities of the poor so that they can participate in the growth of the economy of the country.

Prime Minister Rural Development Fellows Scheme

The Prime Minister Rural Development Fellowship (PMRDF) is a scheme initiated by the Ministry of Rural Development, implemented in collaboration with State Governments. It has dual goals of providing short-term support to the district administration in the underdeveloped and remote areas of the country and develop competent and committed leaders and facilitators who can serve as a resource for a long-term.

National Rural Employment Guarantee Act (NREGA)

As per the National Rural Employment Guarantee Act (NREGA) of 2005, 100 days of employment is guaranteed to any rural household adult who is willing to do unskilled manual work in a financial year. The Act addresses the working people and their fundamental right to live life with dignity. If a person does not get a job within 15 days, he is eligible for getting unemployment allowance. National Rural Employment Guarantee Act (NREGA) also highlights the importance of basic right to work. Amendments have been introduced to this act to minimise corruption in the scheme.

Sampoorna Grameen Rozgar Yojana (SGRY)

The Sampoorna Grameen Rozgar Yojana (SGRY) was launched in 2001 to provide employment to the poor. It also aimed at providing food to people in areas who live below the poverty line and improving their nutritional levels. Other objectives of this Yojana were to provide social and economic assets to the people living in rural areas. The scheme did not include the employment of contractors or middlemen.

Sarv Siksha Abhiyan (SSA)

Pioneered by former Prime Minister Atal Bihari Bajpayee, the Sarv Siksha Abhiyan was launched in 2000. It is an attempt to provide an opportunity to all children between 6 and 14 years of age to get free education which is also a basic fundamental right. The state and the central government share the expenses of this project.

Sansad Adarsh Gram Yojana (SAGY)

Sansad Adarsh Gram Yojana (SAGY) is a rural development project launched in 2014 by the Government of India in which each Member of Parliament will take the responsibility of three villages and look after the personal, human, social, environmental and economic development of the villages. This would substantially improve the standard of living as well as the quality of life in the villages. No fundings have been provided to this project as fundings can be raised through existing schemes.

National Social Assistance Programme

National Social Assistance Programme signifies the fulfilment of Directive Principles in Article 41 and 42 of the constitution which states that it is the duty of the state to provide assistance to the citizens in terms of sickness, unemployment, old age in limits of the economic capabilities. It is basically a centrally sponsored scheme of Government of India which provides financial help to widows, elderly, people with disability in form of pensions. The scheme was launched on August 15 in 1995.

Antyodaya Anna Yojana (AAY)

Launched by the former Prime Minister Atal Bihari Vajpayee in 2000, the Antyodaya Anna Yojana aimed at providing food grains to around 2 crore people at subsidised rates. As per the scheme Below Poverty Line (BPL) families were provided 35 kgs of food grains. Rice was provided at the rate of Rs 3/kg and wheat at the rate of Rs 2/kg. The scheme was first launched in Rajasthan but has now been implemented in all states of the country.

Provision of Urban Amenities In Rural Areas (PURA)

PURA is a strategy for Rural Development in India which was proposed by former President APJ Abdul Kalam in his book Target 3 billion. PURA proposes that urban infrastructure and services should be provided in rural areas to create opportunities outside the cities. This will also prevent the migration of youth from the rural areas to urban areas. The Central Government has been running PURA programs in various states since its launch in 2004.

■■■■

Chapter - 17 Participatory Research Appraisal (PRA)

- PRA has been evolved from RRA (Rapid Rural Appraisal)
- In mid 80's the necessity of participation in rural development became evident and the term PRA was born
- The understanding of PRA came mostly from field rather than academia
- PRA mostly focuses on the empowerment of people through participation
- The sustainability rate of PRA is high due to the participation of the local people
- The sense of ownership and belongingness helps to the success of PRA
- First developed in India and Kenya during the 1980s, it has been mainly used by non-government organizations (NGOs) working on the grass-root level.
- A growing combination of approaches and methods that enable vulnerable people to share, enhance and analyze their knowledge of life and conditions, to plan and act and to monitor and evaluate.
- Participatory rural appraisal (PRA) is an approach used by non-government organizations (NGOs) and other agencies involved in international development. The approach aims to incorporate the knowledge and opinions of rural people in the planning and management of development projects and programmes.
- The philosophical roots of participatory rural appraisal techniques can be traced to activist adult education methods such as those of Paulo Freire and the study clubs of the Antigonish Movement. https://en.wikipedia.org/wiki/Participatory_rural_appraisal - cite_note-4 in this view, an actively involved and empowered local population is essential to successful rural community development. Robert Chambers, a key exponent of PRA, argued that the approach owes much to "the Freirian theme, that poor and exploited people can and should be enabled to analyze their own reality."
- By the early 1980s, there was growing dissatisfaction among development experts with both the reductionism of formal surveys, and the biases of typical field visits. In 1983, Robert Chambers, a Fellow at the Institute of Development Studies (UK), used the term rapid rural appraisal (RRA) to describe techniques that could bring about a "reversal of learning", to learn from rural people directly. Two years later, the first international conference to share experiences relating to RRA was held in Thailand. This was

followed by a rapid acceptance of usage of methods that involved rural people in examining their own problems, setting their own goals and monitoring their own achievements. By the mid 1990s, the term RRA had been replaced by a number of other terms including participatory rural appraisal (PRA) and participatory learning and action (PLA).

- Robert Chambers acknowledged that the significant breakthroughs and innovations that informed the methodology came from community development practitioners in Africa, India and elsewhere. Chambers helped PRA gain acceptance among practitioners. Chambers explained the function of participatory research in PRA as follows:

The central thrusts of the [new] paradigm … are decentralization and empowerment. Decentralization means that resources and discretion are devolved, turning back the inward and upward flows of resources and people. Empowerment means that people, especially poorer people, are enabled to take more control over their lives, and secure a better livelihood with ownership and control of productive assets as one key element. Decentralization and empowerment enable local people to exploit the diverse complexities of their own conditions, and to adapt to rapid change.

Emphasizes empowering local people to assume an active role in analyzing their own living conditions, problems and potentials in order to seek a change in their solution.

An important principle of PRA is to share the results of the analysis between the PRA team and the community members by visualization public presentations and discussions during meetings.

Definition of Participatory Rural Appraisal (PRA)

- It is a participatory method to gather/collect information by involvement of Rural/ local communities for decision making and implementation of the development project, "for the Rural community, by the Rural community and with the Rural Community".
- It is a process to involve the community in planning and decision making.
- Community develop their own skills needed to address issues, analyze options and carry out activities.
- Participatory decision making reflects respect for human dignity and creating the opportunity for individuals to fulfill their responsibility to exercise the right.
- " An approach and methods for learning about rural life and conditions from, with and by rural people"(Chambers, 1994).

Participatory research is a systematic process of involvement of people for the analysis of their own situation and finding out solution for improving their present condition". It can be grouped into two parts.

A. **Explorative research (ER):** is an effort to introduce innovation not known to social system.

B. **Ameliorative research (AR):** It emphasizes in the improvement of what people are doing (at present). It is an effort to make the better compatible.

Participatory Rapid Appraisal is an intensive, systematic but semi-structured leaving experience carried out in a community by a multi-displinary team which includes community members.

The results of the PRA can be used for:

- Need assessment;
- Feasibility studies;
- Identifying priorities for development activities;
- Implementing development activities where new information needs to be collected;
- Monitoring or evaluating development activities.

To conduct PRA will require the following:

- Participation;
- Respect for community members;
- Interest in what they know, say, show and do;
- Patience, without rushing and interrupting;
- Humility;
- Methods which empower community members to express share, enhance and analyze their knowledge.

Main features of PRA

Following are the main features of PRA:

- Triagulation;
- Multi displinary team;
- Mix techniques;
- Flexibility and informality;
- "In the community" approach;
- On the spot analysis;
- Self criticism

PRA : Components

1. Participatory Rural Appraisal Components

- People
- Knowledge
- Participation
- Planning
- Action

2. It is a combination of different approaches to

- Share
- Enhance
- Analyze
- Plan
- Act

3.The secrets behind the success of PRA are

- Decentralization
- Empowerment

Principles of PRA

There are following principles of PRA:

1. **Reversal of learning -** To learn from the local people.
2. **Learning rapidly and progressively -** Exploration, flexible methods, adaptable.
3. **Offsetting bias -** To be receptive rather than preconceived ideas.
4. **Optimizing tradeoffs -** Understanding the usefulness of information.
5. **Triangulating -** Cross checking and approximation.
6. **Empowerment -** The authority to local people through decentralization and confidence building.
7. **Self critical awareness -** Mistakes are lessons to learn and to do better next time.
8. **Personal responsibility -** The belongingness and ownership to the participants.
9. **Sharing -** To discuss and argue about ideas in open forum with all stakeholders.

Process of PRA

- PRA is an approach to project planning and implementation that integrally involves the community through the length of the process.
- Firstly in PRA process villagers or community will assess their needs and to customize the project interventions to their priority concerns and the peculiar circumstances of their community. This will set the objective of the project.
- Then, Participatory mapping is required to be prepared. Participatory mapping is an exercise that uses spatial analysis to gather information about a range of issues and concerns. Community members themselves are asked to do the drawing.
- In drawing participatory maps, the primary concern is not with cartographic accuracy, but rather with gathering useful information that sheds lights on the objectives of the project.

- A meeting should be called at the convenient place with the help of Pradhan/Sarpanch, and all Panchayat members, teachers and any other interested persons, (at least one member of village community should be women) should be invited for the meeting

STEPS OF PRA

- Transects (Systematic walks and observation);
- Informal mapping (Sketch maps drawn on site);
- Diagramming (Seasonal Calendars, flow and causal diagrams, bar charts, chapati diagrams etc.);
- Innovation assessment (scoring and ranking of different actions).

TYPES OF PRA METHODS

1. SPACE MAPPING

- Social and Resource map
- Participatory Modelling Method
- Mobility map
- Services and Opportunities map
- Transect
- Participatory Census Method

2. TIME MAPPING

- Timeline
- Trend Analysis
- Historical transect
- Seasonal Diagram

3. RELATION METHODS

Social effects, Impact, System, network Diagram, Process Map, Well Being, Ranking, Venn Diagram, Pair-wise ranking, Matrix ranking, Pie, Livelihood analysis, Spider Diagram, body Mapping etc.

Characteristics of Participatory Research

Participatory research may be characterized as follows:

1. An activity carried out by group of people to understand their own situation for their own benefit.
2. It is a collective effort to bring desirable results taking into consideration the various aspects of problem in a collective manner.

3. It is an analysis that lead to action.
4. Understanding of personal realities and larger social realities, it will facilitate awareness raising about the connections between micro-aspects of their life and the real macro-aspects of society at large.

It is the involvement of people in decision making, development ownership and control of programmes, products and process.

The success of participatory research lies in if team members of different discipline, function as equal partners with joint responsibility for the final outcome of the research. It is a series of goal achieving acceptable solutions to participant's problem which are linked in circular form by a number of activities.

Over views of Techniques

Over the years techniques and tools have been described in a variety of books and newsletters, or taught at training courses. However, the field has been criticized for lacking a systematic evidence-based methodology.

The basic techniques used are:

- Understanding group dynamics, e.g. through learning contracts, role reversals, feedback sessions
- Surveying and sampling, e.g. transect walks, wealth ranking, social mapping
- Interviewing, e.g. focus group discussions, semi-structured interviews, triangulation
- Community mapping, e.g. Venn diagrams, matrix scoring, ecograms, timelines

To ensure that people are not excluded from participation, these techniques avoid writing wherever possible, relying instead on the tools of oral communication and visual communication such as pictures, symbols, physical objects and group memory. Efforts are made in many projects, however, to build a bridge to formal literacy; for example by teaching people how to sign their names or recognize their signatures. Often developing communities are reluctant to permit invasive audio-visual recording.

Developmental Changes in PRA

Since the early 21st century, some practitioners have replaced PRA with the standardized model of community-based participatory research (CBPR) or with participatory action research (PAR). Social survey techniques have also changed during this period, including greater use of information technology such as fuzzy cognitive maps, e-participation, telepresence, social network analysis, topic models, geographic information systems (GIS) and interactive multimedia.

Cyclic Process of Participatory Research

Participatory research is a cyclic process starting from grass-root (practioners) to international level scientists. In this process, all those are involved, who equally participate in making the programme, executing and getting benefit out of it.

This model comes together at the point of intersection and shows that the demand for participation will effect the support required and delivery system will be conditioned by the materials provided and so on. The limitation of the present model may be the last participation of practitioners in various systems. There is need to consider the importance of participatory research at various levels of society. In this process all those are involved, who equally participate in making the programme, executing and getting benefit. It is group activity.

Following are the various steps in conducting a participatory research:

1. **To know the problem:** All the members of the group should first agree on a common statement of a problem. Some common problems, for example, low income, indebtness, sickness of children etc. The problem should be stated clearly and concretely.
2. **Solution is needed or not?** : Sometimes it happens that we have the problems but we are not interested in its solution at a time. We do not make any efforts for its solution. This may happen in two situations.
   a. When we know that solution is within our control and it can be solved at any time, or
   b. When the problem is beyond our control and we think that it is wastage of resources and time because we can not solve it.

   Therefore, the group should express interest in solving the problem.
3. **To know the present status and additional information needed for the problem to be tackled:** All previous information (history) known to the group should be recorded about the problem. For example, if the problem is sickness of children, then they can each describe the sickness, its type, intensity and duration in the case of their own children. This helps in focusing on the problem in a concrete way.
4. **The ways of collection of additional information:** This involves the steps of data collection. Who get the information, whose assistance is needed, who will for what and upto what time the information will be available?
5. **Analyzing the available information:** Having obtained the information the group collectively analyses that information. It is important to note that analysis and reflection is undertaken as a collective process, not to be done by one person alone. This will help the group to understand, the possible causes and provide clues for solution.
6. **What actions are needed for possible solutions:** Various solutions to the problem are sorted out, their workability is assessed and the choice of solution is made. Then, for solving the problem a decision has to be made that what actions are to be taken, by whom, when, where and how. After this, actions are taken.
7. **Review of the actions:** Systematic assessment of the impact of the action taken should be made to see if the problem is fully or partially solved. The actions taken may sometimes create new problems. For example, if in a community there is need for safe drinking of water and tap water is provided to solve this problem, the problem may create another problem of drainage in the lanes. Now, a group exercise will be needed to work out agreed solution for creating a drainage system in the community.

Generating Acceptable Technology (GAT model)

The involvement of interdisciplinary teamwork in all phases of ongoing research and its transfer for the use of practioners at the grassroot level. It is based on the assumption that the outcome of research should be applied and transfer should begin and end with the community. For official utilization of available resources a proper assessment of problem is essential. The participants made a final decision after evaluation of technology under their own situation, using their own resources and management. The scientists monitor the technology time to time to determine how it is beneficial for participants and consumers.

For the best use of available resources, proper diagnosis of the problem is most important step. It may require the help of research scientist to make in depend observation. The next step becomes very important to sort out the common problems of the participants for which solutions are to be evolved. In the process of defining the common problem, interdisciplinary dialogue and interaction, generally initiated by debate, taking into consideration the socio-technical perspectives together, constructive suggestions are noted.

In this stage, community may put forth some solutions by itself that should also be thoroughly examined and if only improvement in the same is needed the possibility should be workedout. The research team should make all efforts to evolve some potential solution of the problems. It should be tested in the community situation for adoption. When the solutions are worked out, it should be finally evaluated by the scientist in cooperation with participants under the existing resources and management. Even at this stage the community may accept or reject the technology. If rejected, it gives further path for research to find out the reasons of rejection and may find ways to improve the performance and acceptability of the technology. If accepted the scientist monitor time to time to determine how it is beneficial to other communities.

Tools and techniques of PRA

Direct Observation – is systematically observing objects, events, processes, relationships or people and recording these observations. Direct observation is a good way to cross-check respondent's answer. Use a checklist to do observations systematically. The steps are:

- Think about the objectives and broad topics of the PRA;
- Identify indicators which you can asses through direct observation;
- List down the indicators.

Review of Secondary Data Sources - secondary data sources are sources of information that are relevant to the area or subject of the planned PRA and are available in published or unpublished form.

Construction of Diagrams - a diagram is any simple model which presents information in an easily understandable visual form. Diagrams are constructed to:

- Greatly simplify complex information;
- Show/picture an analytical procedure;
- Facilitate communication;
- Stimulate discussion;

- Increase consensus among members;
- Involve community members and to discover their views and categories by encouraging them to draw diagrams of their own.

Seasonal Calendar- a calendar showing the main activities, problems and opportunities throughout the annual cycle in diagrammatic form (multi-layer diagrams in one sheet). It helps identify the events that have impart on the people's lives. Information in the calendar include dates, events, cashflow and even input prices.

Semi-structured Interviews- is a guided interview where only some of the questions are predetermined and new questions come up during the interview. The interviewer prepares a list of topics and questions rather than a fixed questionnaire. These targets are individuals, key informants and focus groups.

Time Trends – are qualitative changes over time and can be used for many variables (those that have impact to the lives of the people on the area). Examples are input prices, rainfall, livestock population etc.

Historical Profile – chronological list of events in the community. This reveals important information for understanding the present situation in a community.

Daily Routine/Daily Activity Profile - collection of day – to – day activities of community members.

Flow Diagram – this diagram shows key institutions and individuals in a community and their relationships and importance for decision making.

Drawing Maps and taking Photographs – draw maps of the area are sources of spatial information in the area, and taking photos will show the features of the area. An ideal method for mapping is to conduct "participatory mapping". While "transect mapping" is mapping of the main land use zones of the area (includes soil, land use, crops and vegetation), problems and opportunities – presented in linear distance along the transect path.

The Maps Draw and Photos taken may be done for topics such as:

- Demography
- Social and residential stratification
- Use of natural resources
- Fields and land use
- Spatial arrangement of houses and use of space
- Mobility
- Water
- Soils

Analysis Group Discussion – also the focus group discussion where, the participants are allowed to conduct a free discussion on important and related topics. Rules for analysis group discussion are:

- Listen

- Learn
- Facilitate
- Don't dominate
- Don't lecture
- Don't interrupt
- Respects people opinions

Ranking – ranking or scoring is placing something in order. This complements semi-structured interviewing by generating basic information which leads to more direct questioning.

- Preference ranking
- Pairwise ranking
- Direct matrix ranking
- Wealth ranking

Innovation Assessment – helps to assess and prioritizes possible options for development activities. It also identifies the innovation (intervention) and its implementation. This activity is done after the analysis group discussion.

Sustainability Analysis- this is useful for the periodic review of development activities. Without collection of information, this sharpens the group's analytical skills and brings out important issues. The questions to be answered are:

- Impact indicators
- Goals/objectives
- Output indicators
- What should we continue doing
- Strengths- weakness
- What should we stop doing
- What should we start doing

Use of Media in Participatory Research

The rational underlying the use of communication to enhance popular participation is that people can only be motivated towards decision making. If firstly, that they have been consulted by the planners during the design phase of a development project in other words, their felt needs and aspirants have been taken into accounts and secondly, if they have information on which to base their decision. They also need a continuing possibility to making their opinion known as the project proceeds.

Communication techniques using such devices as CD/DVD/Computer or audio tape recorders, photography, rural newspapers and the like can give life to a consultation process through a two way flow of information. They can also provide, in a comprehensible, audience tailored form, the information needed for decision making. Even TV and radio, normally one way channels, by imaginative use of audio and video recordings made in the field, among the target audience and by subsequent broadcasting to them and to peer groups can be used effectively.

Sometimes the group engaged in Participatory Research may find itself uncomfortable with written methods. In such circumstances, simple audio-visual methods can be used to facilitate analysis and reflection. A group discussion or individual interview can be taped and then can be replayed for analysis.

Women can be encouraged to draw their perceptions of the problem on a blackboard, paper or cloth. Black and white photographs, cutting from old magazines and other materials used to promote reflection and analysis. These methods can easily be combined with other methods to express the perception of their problems such as PRA.

It is important to recognize that the methods listed below represent the range of ways in which participatory research can be conducted. They can also be used in combination, and new methods can also be invented based on a given situation.

1. **Small survey:** A well known method of data collection is to carry out a small survey. A group of people can investigate a particular problem by using questionnaire to know the opinion of a larger number of people. The questionnaire should be simple and have only a few (8-10) questions. Each questions should be short, clear and straight forward. The questionnaire should seem like an easy conversation. The information obtained may be tabulated and analysed to know the response of majority of the people about a particular problem.

2. **Group discussion:** A small number of people (10-25) come together to share experience and decide for any action through group discussion on a particular problem. It is one of the most popular and easy method of combining data-collection, analysis and action planning. In a group discussion, contradictory opinions and points of view of various members can be openly discussed and analyzed. This facilitated a deeper problems and analysis of the problems as well.

   The important considerations of group discussion is to insure the comfortable atmosphere in the group so that all may participate actively and listen to each other. If the group is large, it can easily be subdivided into smaller groups to facilitates free an easy discussion. This method can easily be combined with other methods.

3. **People's camp:** A camp or shivir of workshop or training programme of people can be used an important event to carry out participatory research. The period of the camp may be put over a day or two with common lodging and boarding arrangements. Generally, people from different villages, over a long distance come together.

   In addition to education of people, it can provide an informal and free atmosphere. In a camp approximately 50-100 people can participate easily depending upon the facilities available at a point of time and place.

4. **Research team:** For many problems chosen by the group, they may not have the amicable solution and it may require the help of scientists for technical advice. The scientists with their team extend their services and expertise to cater the needs of the group.

5. **Visit to a demonstration :** It is always beneficial to visit an area where a similar problem has been solved for which the group is interested. It helps to feel confident. After the visit, this can be shared with other members of the community to solve their problems.

6. **Folk theatre:** Indian folk theatre has an important role to play in the process of social change in India. The folk theatre media, low cost and indigenous and may prove better for focusing and solving rural problems. Rural poor can communicate better and convince people by making their own theatre.

   Indian folk theatre is full of exploitative stories of kings and priests tuned into Gods, picked up from different Hindu epics like Ramayana and Mahabharat. In some places, simply produced dramas have been used as an effective method of participatory research.

7. **Puppets:** Puppets refer to any inanimate thing animated on stage to communicate any subject. All the six instincts (Communicative, dramatic, artists, musical, inquisitive and constructive) basic to human learning are stimulated and get satisfied through puppet show. The puppetry is a powerful audio-visual media for group communication. It is also a folk and people's art forum. Puppet can be made out of easily available and cheap materials in the community like peer match, cloth, cardboard, mud, saw dust, hay etc.

   In addition to all above points, puppetry is an "INDIAN" medium of communication which attracts people of all age groups. In this way, the puppets can be used by the local communities for highlighting their problems and related perceptions. It can be used to communicate the possible solutions also.

8. **Documents:** Many books, reports, papers, registers, diaries and documents may contain useful information about a particular problem. Analysis of these documents is another method which may be of some use to solve the problems of people.

9. **Use of video:** Video is most suitable medium for generating interaction. It has the ability to present the community in both moving image and sound for immediate reproduction. Video tapes can be used for promoting audiovisuals literacy for motivation, attitudinal change, behaviour reinforcement, community participation and entertainment. Similar results even with more clarity an immediacy can be now used through computer system.

■■■■

Chapter - 18 Agricultural Extension Systems in India

India has made significant achievement in agriculture for increasing food production by four folds during last six decades. Public extension played a major role in ushering green revolution in Indian agriculture. However, considering the varied agro-ecological situations under which farmers operate besides variations in the resource base of farmers, the extension system envisaged to achieve desired growth in agricultural sector has to be pluralistic in nature and hence multiplicity of extension systems are operating in India.

The Union Government mainly provides road map through its policies, programmes and budgetary support to the sector. The programmes conceived at national level are mainly implemented by the states through its development departments. Besides, states also formulate region specific development programmes.

Similarly, Indian Council of Agricultural Research (ICAR) is an apex body at the national level that supports research and extension activities to evolve effective Transfer of Technology (TOT) models. The State Agricultural Universities also contemplates to develop extension models suitable to take up transfer of technology besides implementing the models evolved by ICAR system.

India has witnessed and experienced many changes in extension approaches before and after independence for the development of agrarian society.

Before Independence

Only some sporadic attempts by individuals, political and social leaders and philanthropists were made in different parts of the country, the most remarkable ones being Shriniketan Project (1921) by Rabindranath Tagore, Marthandam Project (1921) and Firka Development scheme in Madras state (1946). These projects were local specific and aimed at holistic rural development but gradually disappeared with time due to lack of government support and mass involvement. The projects were basically around individual philosophies towards rural development.

During 1950s

- The central task of extension was to help rural families help themselves by applying crude knowledge of farming, homemaking, and family and community living.

- Notable Extension Programmes in India during the period- Community Development Programme (CDP), National Extension Service (NES).

During 1960s

- Emphasis was started laying on 'rural education'- Agricultural extension became a system of 'out-of-school education' for rural people.
- Extension personnel had the task of bringing scientific knowledge to farm families in the farms and homes. The object of the task was to improve the efficiency of agriculture.
- Notable Extension Programmes in India during the period- Integrated Agriculture Development Programme (IADP), High Yielding Variety Programme (HYVP).

During 1970s

- Extension was a service or system which assisted farm people, through educational procedures, in improving farming methods and techniques, increasing production efficiency and income, bettering their standard of living and lifting social and educational standards.
- Extension involved the conscious use of communication of information to help people form sound opinions and make good decisions.
- Notable Extension Programmes in India during the period- Krishi Vigyan Kendra (KVK), Training and Visit (T &V).

During 1980s

- Introduction of centrally sponsored rural employment generation programmes.
- Emphasis laid upon rural housing and rural infrastructure.
- Extension became a professional communication intervention deployed by an institution to induce change in voluntary behaviours with a presumed public or collective utility.
- Extension Programmes in India during the period- Integrated Rural Development Programme (IRDP), National Rural Employment Programme (NREP), Jawahar Rojgar Yojana (JRY).

During 1990s

- Added thrust upon rural employment generation and self employment.
- Extension perceived as the organized exchange of information and executed as deliberate transfer of skills.
- The essence of agricultural extension was to facilitate interplay and nurture synergies within a total information system involving agricultural research, agricultural education and a vast complex of information-providing businesses.

- Extension Programmes in India during the period- Jawahar Gram Samriddhi Yojana (JGSY), Swarnajayanti Gram Swarojgar Yojana (SGSY), National Agricultural Technology Programme (NATP), Institute Village Linkage Programme (IVLP).

2000s onwards

- Reshaping extension into pluralistic mode.
- Added thrust upon Research-Extension linkage.
- Structured initiatives in Public-Private-Partnership (P-P-P).
- Dependence on use of Information Communication Technology (ICT) tools.
- Extension Programmes in India during the period- National Agricultural Innovation Project (NAIP), Agricultural Technology Management Agency (ATMA), Kishan Call Centre (KCC), Agri-Clinic and Agri-Business Centre (ACABC), NFSM.

Post-independence period

Immediately after independence, India faced two major challenges on food security fronts-

- the grain bowls of west Punjab and East Bengal went to Pakistan and
- mass-exodus of people from across the border.

The gravity of the situation compelled the Government of India to lodge a programme known as 'Grow More Food Campaign'. The first 3 more or less structured effort towards holistic rural development was the community development programme (1952) followed by the national extension service (1953). In the sixties, agriculture production situation was very critical and there was a felt need for intensification of agriculture with the use of high yielding varieties. Majority of the population being dependent solely upon agriculture for earning their livelihood agricultural development was the main indicator of rural as well as economic development of the country. A number of programmes, namely Integrated Agriculture Development Programme (IADP), Intensive Agriculture Area Programme (IAAP), National Demonstration (ND) and High Yielding Variety Programme (HYVP) were launched. The approach although produced immediate results, did not succeed in long run due to its failure to uplift the small and marginal farmers and in terms of uniformly reaching every pockets of the country. A programme, namely Integrated Rural Development Programme (IRDP) launched during the early eighties was more or less a broadened approach towards rural development. As far as agricultural extension in specific is concerned, the Training and Visit (T&V) system launched during the seventies was a breakthrough approach which laid major emphasis on trained field extension agents in technology transfer. The importance of disseminating the research based recommendations was in real sense for the first time realized through this approach. Using trained field staff for technology transfer was subsequently executed through the Krishi Vigyan Kendra (KVKs) project in a more systematic manner. The new approach was found to have immense potential not only in mere transfer but internalization of technologies as it encouraged education and peoples' participation. The approach with a number of modifications has been found sustainable and is under operation all over the country.

The first planned attempt started with the launching of Community Development Programme in 1952, followed by the National Extension Service in 1953. These programs were able to educate responsive farmers to take up improved methods of farming across the country. The other important Area-Based Special Programmes were; Intensive Agricultural District Programme (IADP, 1960), Intensive Agriculture Area Programme (IAAP, 1964) and High Yielding Varieties Programme (HYVP, 1966) besides Farmers Training Centres (1967) to train farmers on high yielding varieties and improved methods of farming to back up the above programs. The cumulative effect of these programs resulted in increased productivity, which made way to usher in 'Green Revolution' in Indian agriculture during late 1970s.

However, these programs widened gap between resource rich and resource poor farmers. In order to enable resource poor farmers to take benefit of improved farm technology, many client-based programs were introduced. The most important ones being; Small Farmers Development Agency (SFDA, 1969), Marginal Farmers and Agricultural Labourers Programme (MFAL, 1969), District Rural Development Agency/Society (DRDA, 1976), Integrated Rural Development Programme (IRDP, 1978) and Lab to Land Programme sponsored by ICAR (LLP, 1979). Although, these programs were able to improve the socio-economic conditions of beneficiaries, they were isolated in a given time and implemented in a phased manner.

By the middle of 1980s it was observed that extension services in the developing countries were suffering from a number of weaknesses, including the dissipation of extension workers' energies on low priority tasks; the lack of single as well as clear line of command; and low level of agricultural knowledge and skill among field level functionaries. As a means of reforming and strengthening the extension service, a reorganized agricultural extension system known as 'Training and Visit' (T & V) system was introduced in the country.

Training and Visit System

This system was introduced in India in 1974 with the World Bank assistance. Training and Visit system became the dominant method of restructuring the extension services in over sixty countries in Asia, Africa and Latin America. The system aimed to achieve change in production technologies of farmers through professional assistance for the contact farmers from well trained extension personnel on agricultural research and supported by supply, service and marketing facilities which were earlier lacking in National Extension Service.

The important features of T & V system are : (1) Professionalism, (2) Single line of command, (3) Concentration of efforts, (4) Time-bound work, (5) Field and farmer orientation, (6) Regular and continuous training and (7) Linkage with research.

Training

Training was imparted to the extension personnel at two levels; in the first level the middle level officers (SMSs) of the department of agriculture was trained by the university scientists (master trainers) in the monthly workshops generally organized at the district level in the nearby research station. Subsequently (second level) the SMSs used to impart training to grass root level extension functionaries in the fortnightly meetings in the respective talukas. Over the period, strategic changes were made in the conduct of training to meet growing demand of technology and at present bimonthly workshops are being conducted in majority of the states.

Visit

Knowledge and skill gained by the field extension functionaries during the fortnightly training programme was transferred to contact farmers through regular field visits. These contact farmers were in turn shared the knowledge and skill to 10- 15 fellow farmers.

The main weakness of T&V system was it largely remained as Departmental program characterized by centralized operations with inadequate farmer driven and farmer accountable feedback systems resulting in low level of farmers' involvement. The single-discipline-commodity oriented approach had serious limitations in rainfed areas where holistic farming systems approach was required. This further lacked focus on location specific needs of regions, disadvantaged areas and target groups.

Broad Based Extension System (BBES)

This system aimed at rectifying the defects of T & V system in some of the states. In the BBE :

(a) the role of subject matter specialists was amplified and they were invited to formulate messages suitable to their land based activities (agriculture, sericulture, animal husbandry, horticulture, Human resource development, creating agriculture infrastructure etc.).

(b) village extension workers had full time job by offering messages during lean season also.

(c) the concept of broad based education laid emphasis on formulating and delivering composite messages to the farmers to meet the needs of their full agricultural environment.

Agriculture Technology Management Agency (ATMA)

In India where Agro-climatic zones widely differ besides significant variation in socio-economic status of farmers', uniform extension service is not the panacea for all the regions. It was realized that public extension system will have to be placed in new decentralized institutional arrangements which are demand driven, farmer-accountable, bottom-up and have farming system approach. To address these issues, the ATMA was envisaged as alternate public extension institution.

ATMA extension mechanism/model was firstly implemented in Andhra Pradesh, Bihar, Himachal Pradesh, Jharkand, Maharastra, Orissa and Punjab covering 4 districts in each state during 1998 under the guidance of National Institute of Agricultural Extension Management (MANAGE), Hyderabad. The evaluation report of Indian Institute of Management (IIM) Lucknow revealed that the ATMA's extension approaches have been proving to be very promising in execution of the reforms and thus the progress was extended to other states.

Government of India is funding ATMA programme in all the districts of the country. ATMA is a registered society of key stakeholders (farmers, line/development departments, non government organizations, input dealers, mass media, agri-business companies, farmers organizations, etc.) involved in agriculture activities for sustainable agricultural development in the district. Though the State Department of Agriculture serves as a nodal agency for implementing ATMA, the programme aims to increase coordination and integration among developmental departments. Emphasis has been laid on providing flexible working environment and establish effective integration of all the stakeholders at the district level and thereby increase input in to programme planning and resource allocation, especially at the block level and thereby increase accountability of stakeholders.

Every district has to prepare the Strategic Research and Extension Plan (SREP) for implementing ATMA in respective districts. The SREP is prepared through participatory methodologies such as Participatory Appraisal Techniques involving all the stake holders and farmers. The SREP contains detailed analysis of all the information on existing farming systems in the district and research-extension gaps required to be filled up. It also prioritizes the research-extension strategies within the district. It becomes the basis for development work plans at block/ district level. ATMA is a more comprehensive farmer centric bottom-up approach extension program which is in operation in all the districts of the country.

Extension approach of State Development Departments

Agricultural development being the state subject, the major responsibility of implementing all the programmes of central and state governments goes with state department of agriculture. The major State Development Departments (Departments of Agriculture, Horticulture, Sericulture, Animal husbandry and Fisheries) having focus on development of agriculture and allied activities to undertake outreach activities. The effective extension system is visible only in the case of Agriculture Department, which was one of the major partners in implementing Training and Visit System. Some of the states have suitably modified T and V system (like Broad Based Extension System in Tamil Nadu) to suit their local needs and also incorporate Demonstration as one of the components of the programme.

Presently the State Department of Agriculture is a nodal agency to implement Agriculture Technology Management Agency (funded by Ministry of Agriculture and Cooperation, GOI) at the district level. The coordination and integration of activities of other development departments is envisaged under ATMA.

4H-Club

- 4H-Club was established in 1990 in America to provide training to youth for agricultural and home development, it was first Government step.
- 4H-Club means Heart, Head, Hand and Health was explained by A.B. Graham.
- On the basis of 4H – Club of USA, young farmer's association of India was started by P.S. Deshmukh in 1956.

Support to State Extension Programme for Extension Reforms

Support to State Extension Programme for Extension Reform was launched in May 2005 in 267 districts of all the state/UTs in the country. Centrally sponsored "Extension Reforms (ATMA) Scheme" is based on new methodological and institutional arrangement from state headquarter to village level. Top priority has been given to farming System Approach and Decentralized Planning and Decision making process at various levels. Under this scheme extension efforts are focused community based farmer's group. Government of India is providing necessary guidance, logistic support and valuable technical and financial existence to revamp and revitalize the prevailing extension system through many strategic efforts including innovation in technology dissemination, validation of ITKs, encouraging PPP, involvement of NGOs, FOs, Agri. Preneurs and PRIs etc. linking of production system with markets and revision of cafetaria of activities in accordance with the emerging needs of the extension system. During 12th Five Year Plan the name of ATMA was changed to National Mission of Agricultural Extension and Technology.

National Food Security Mission (NFSM)

The National Development Council (NDC) adopted a resolution to launch a food security mission comprising rice, wheat and pulses to increase the production of rice to 10 million tons, wheat by 8 million tons and pulses by 2 million tons by the end of eleventh plan (2011-2012). Initially this project was started in 17 states of the country. A centrally sponsored scheme, NFSM has been launched from October, 2007. Accordingly, a centrally sponsored scheme, 'National Food Security Mission' has been launched from 2007-08 to operationalize the above mentioned resolution.

The National Food Security Mission' initially have three components

i. National Food Security Mission – Rice

ii. National Food Security Mission-Wheat

iii. National Food Security Mission-Pulses

During 2018-19, three more components have been included

i. National Food Security Mission-Coarse cereals

ii. National Food Security Mission-Nutri cereals

iii. National Food Security Mission- Oilseeds

Objectives

NFSM have following objectives:

1. Increasing production of rice, wheat and pulses through area expansion and productivity enhancement in a sustainable manner in the identified districts of the country.
2. Restoring soil fertility and productivity at the individual farm level
3. Creation of employment opportunities.
4. Enhancement farm level economy (*i.e.* farm profit) to restore confidence amongst the farmers.

At the district level, the scheme was implemented through the Governing Board of Agricultural Technology Management Agency (ATMA). The state level agency provided the required funds to the district level agency for the execution of the programme at the district/block level. A district food security mission Executive Committee (DFSMEC) was constituted for project formulation, implementation and monitoring of the scheme components through the District Agricultural Department. The project management team was assisted by technical experts appointed on contractual basis for technology transfer and implementation of the mission.

KVK's, SAU's and ICAR institutes functioning in the district provided technical support in formulation of projects, its implementation and monitoring. The state level agency would ensures implementing of the programme, in a time bound manner and could make funds available to the District level agency in accordance with their approved programme. Funds was released in installments based on the progress report and submission of utilization certificates.

More decentralized Extension network- Raitha Samparka Kendra (RSK)

In order to have more decentralized extension network below the district level for effective dissemination of modern agricultural technologies, Government of Karnataka has launched "Raitha Samparka Kendra (RSK)" during 2000, with the objective of providing extension services to the farmers at Hobli level. Presently there are over 745 RSKs located at Hobli level functioning under the administrative control of Zilla Panchayat. The main objectives of the RSKs include:

(a) providing updated information on crop production and marketing
(b) facilitate on the spot provision of agricultural inputs,
(c) facilitate on the spot provision of testing the agricultural materials, and
(d) provide a forum of interface with public and private technologies and inputs.

The State Department of Horticulture is implementing National Horticulture Mission since 2000. The Human Resource Development through training and demonstrations is an integral part of the mission. Under this programme, training of the farmers, field level workers and officers by both Government sectors (SAUs and ICAR Institutes) and Non-government Organizations is being undertaken.

Departments of Sericulture, Animal husbandry and Fisheries are conducting a variety of extension activities like discussion, meetings, demonstrations, field visits, fairs, field days etc. by the technical field functionaries to transfer the new technologies.

Initiatives of ICAR

The Indian Council of Agricultural Research (ICAR) took up number of extension programmes over the years. The first programme was National Demonstration Scheme (1964) initiated during 1964-65 to demonstrate the production potentiality of major crops in the farmers' field. The Operational Research Projects (ORPs) was started during 1975 to identify technological as well as socio-economic constraints and to formulate and implement the problem solving technology modules on area/watershed/ target group basis in operational area. Lab-to-Land programme was launched during 1979 to transfer low cost technologies in agriculture and allied enterprises. As part of technology mission on oilseeds and pulses, the council started Frontline demonstration in 1990-91. Institution –Village Linkage Programme was 6 launched during 1995 which was funded under National Agricultural Technology Project (NATP) during 1998-2004. Under innovation technology demonstration component of NATP, the ICAR established Agriculture-Technology Information Centre (ATIC) in State Agricultural Universities to work as single window support system linking the various units of research institution with intermediary users and farmers in decision making and problem solving exercise through availability of technology inputs, products, information and advisory services under one roof. Since 2006-07, ICAR is implementing National Agricultural Innovation Project (NAIP), in a consortium mode.

Krishi Vigyan Kendra (KVK)

Krishi Vigyan Kendra (Farm Science Centre) is an innovative institution of ICAR established at district level. The first KVK was established at Pondicherry during 1974 and has grown as a largest network in the country with 716 KVKs during January 2020. KVKs are 100 per cent funded by ICAR and administered by ICAR institutes/SAUs/Deemed Universities/Non-government Organizations or State Department of Agriculture.

KVKs play a vital role in conducting on farm testings to identify location specific agricultural technologies and demonstrating the production potential of crops at farmers' fields through frontline demonstrations. They also conduct need based training programmes for the benefit of farmers and farm women, rural youths and extension personnel to update their knowledge and skills and to orient them in the frontier areas of technology development.

KVKs are creating awareness about improved agricultural technologies through large number of extension programmes. Critical and quality inputs like seeds, planting materials, organic products, biofertilizers and livestock, piglet and poultry strains are produced by the KVKs and made available to the farmers. Agricultural Knowledge and Resource Centres are set up at KVK to support the initiatives of public, private and voluntary sectors at district level. A number of successful case studies have emerged out of effective implementation of various technological and institutional interventions by KVKs. The KVKs are evolving as the future grass root level institutions for empowering the farming community. KVKs have made dent and has become part of decentralized planning and implementation instrument to achieve desired level of growth in agriculture and allied sector. Studies were conducted from both internal and external agencies on performance of KVKs at different point of times have indicated that the KVKs have significantly contributed in educating farmers in improved practices and enhancing productivity levels.

State Agricultural Universities (SAUs)

State Agricultural Universities apart from lending support in implementing ICAR sponsored extension programmes, have evolved several innovative extension models to effectively reach the farming community. The type of extension activities undertaken by SAUs vary from state to state. The UAS Bangalore is undertaking transfer of technologies to farmers and others through Krishi Vigyan Kendras, Extension Education Units, Farmers Training Institute, Bakery Training Unit, Staff Training Unit, Agriculture Technology Information Centre by organizing diversified extension educational activities *viz.*, farm trials, demonstration, meetings, discussion, conventions, training programme, farmers field school, field days, krishi mela, exhibition, agricultural campaign, educational tours, exposure visits, diagnostics visits , farm advisory services etc. The SAUs publish agriculture literature (books, package of practices, booklets, folders, and leaflets) in local languages for dissemination of agricultural technologies to the farmers and extension personnel. Expert Centres and Village Resources Centres are established in collaboration with Indian Space Research Organisation (ISRO) for interaction of farmers with experts on a variety of agricultural information. Mobile message services and Kissan Call Centre (established at ATIC) as well as KVKs are providing timely information to the farming community regarding agricultural technologies, weather data and market information.

Extension activities of Commodity Boards, Financial Institutions, Input agencies, Non-Government Organizations and Media Organizations

Commodity Boards (Coffee board, Spice Board, Tobacco board, Groundnut board, Coconut development board etc.) are extending crop/ commodity specific technical know-how to the farmers to a limited extent as many of these boards do not have grass root level functionaries throughout the country. Financial institutions normally provide assistance in preparation of agriculture project proposals by their technical staff to the farmers and others. Agricultural input agencies besides providing critical inputs like seeds, planting materials, fertilizes, plant protection chemicals etc, they also sponsor/organize training programme to educate farming community. The Media organizations (Print and electronic media) are disseminating timely information on weather, technical information and marketing information. Various committed Nongovernmental organizations and Philanthropists are also rendering rural extension services to the rural community in the field of agriculture and allied sectors, health, sanitation, education, water supply etc., across the country.

Private Extension Strategies

The private players have emerged strong and competitive with huge capital investment and manpower to extend their reach to the unreached. The following efforts taken by some private extension agencies may be exemplified as alternate extension models:

The 'Tata Kishan Sansar' (TKS) model of Tata Chemicals Ltd.

- Providing need based operation and advisory support to the farmers.
- A franchise based 'hub and spoke' model that provides services like remote diagnosis, soil testing, house brands for cattle feed, sprayer, seed, pesticides etc.
- There are now 32 hubs catering to 681 TKSs covering around 2.7 million farmers, in 22,000 villages of 88 rural districts of the country.

The 'Hariyali Kishan Bazar' model of the DCM Shriram Consolidated Ltd.

- Business model providing 'end-to-end' agri-business solution to the farmers.
- A Package of agri-input, extension, credit and produce-marketing.
- Operates in a chain of more than 300 rural retail stores across eight states, catering to 45, 00,000 farmers.
- Emerged as 'rural super bazar'.

The 'Mahindra Krishi Vihar' model of Mahindra and Mahindra Ltd. - 'One stop farmer solution shop'

An innovative and successful Extension strategy of an NGO, Digital Green

- Started functioning in 2008.

- Dissemination of information to small and marginal farmers by means of 'digital video'.
- Delivery of targeted contents those are scalable to large number of farmers.
- Produced more than 1200 such videos on diverse farming technologies and practices, disseminated by face to face contact and uploaded in their website.

Pradhan Mantri Fasal Bima Yojana

The **Pradhan Mantri Fasal Bima Yojana (PMFBY)** launched on 18 February 2016 by Prime Minister Narendra Modi is an insurance service for fanners for their yields. It was formulated in line with One Nation - One Scheme theme by replacing earlier two schemes **National Agricultural Insurance Scheme (NAIS)** and **Modified National Agricultural Insurance Scheme (MNAIS)** by incorporating their best features and removing their inherent drawbacks (shortcomings). It aims to reduce the premium burden on farmers and ensure early settlement of crop assurance claim for the full insured sum.

PMFBY aims to provide a comprehensive insurance cover against failure of the crop thus helping in stabilising the income of the farmers. The Scheme covers all Food & Oilseeds crops and Annual Commercial/Horticultural Crops for which past yield data is available and for which requisite number of Crop Cutting Experiments (CCEs) are being conducted under General Crop Estimation Survey (GCES). The scheme is implemented by empanelled general insurance companies. Selection of Implementing Agency (IA) is done by the concerned State Government through bidding. The scheme is compulsory for loanee farmers availing Crop Loan /KCC account for notified crops and voluntary for other others. The scheme is being administered by Ministry of Agriculture.

The scheme has been beset by a number of problems for the farmers with unpaid dues in thousands of crores while insurance companies have pocketed the money.

Objective of the Scheme

Pradhan Mantri Fasal Bima Yojana (PMFBY) aims at supporting sustainable production in agriculture sector by way of

- Providing financial support to farmers suffering crop loss/damage arising out of unforeseen events.
- Stabilizing the income of farmers to ensure their continuance in farming
- Encouraging farmers to adopt innovative and modern agricultural practices
- Ensuring flow of credit to the agriculture sector which will contribute to food security, crop diversification and enhancing growth and competitiveness of agriculture sector besides protecting farmers from production risks.

Implementing Agency

The Scheme shall be implemented through a multi-agency framework by selected insurance companies under the overall guidance & control of the Department of Agriculture, Cooperation &

Farmers Welfare (DAC&FW), Ministry of Agriculture & Farmers Welfare (MoA&FW), Government of India (GOI) and the concerned State in co-ordination with various other agencies; *viz.* Financial Institutions like Commercial Banks, Co-operative Banks, Regional Rural Banks and their regulatory bodies, Government Departments viz. Agriculture, Co-operation, Horticulture, Statistics, Revenue, Information/Science & Technology, Panchayati Raj etc.

Coverage of Farmers

All the farmers growing notified crops in a notified area during the season who have insurable interest in the crop are eligible.

Compulsory coverage: The enrollment under the scheme, subject to possession of insurable interest on the cultivation of the notified crop in the notified area, shall be compulsory for following categories of farmers:

- Farmers in the notified area who possess a Crop Loan account/KCC account (called as Loanee Fanners) to whom credit limit is sanctioned/renewed for the notified crop during the crop season.
- Such other farmers whom the Government may decide to include from time to time.

Voluntary coverage : Voluntary coverage may be obtained by all farmers not covered above, including Crop KCC/Crop Loan Account holders whose credit limit is not renewed.

Coverage of Crops

1. Oil seeds
2. Food crop
3. Annual Commercial / Annual Horticultural crops.

In addition for perennial crops, pilots for coverage can be taken for those perennial horticultural crops for which standard methodology for yield estimation is available. A

Coverage of Risks and Exclusions

Following stages of the crop risks leading to crop loss are covered under the Scheme.

1. **Prevented Sowing/Planting/Germination Risk:** Insured area is prevented from sowing/planting/ germination due to deficit rainfall or adverse seasonal/weather conditions.
2. **Standing Crop (Sowing to Harvesting):** Comprehensive risk insurance is provided to cover yield losses due to non-preventable risks, *viz.* Drought, Dry spell, Flood, Inundation, widespread Pests and Disease attack, Landslides, Fire due to natural causes, Lightening, Storm, Hailstorm and Cyclone.
3. **Post-Harvest Losses:** Coverage is available only up to a maximum period of two weeks from harvesting, for those crops which are required to be dried in cut and spread/small bundled condition in the field after harvesting against specific perils of Hailstorm, Cyclone, Cyclonic rains and Unseasonal rains

4. **Localized Calamities:** Loss/damage to notified insured crops resulting from occurrence of identified localized risks of Hailstorm, Landslide, Inundation, Cloud burst and Natural fire due to lightening affecting isolated farms in the notified area.
5. **Add-on coverage for crop loss due to attack by wild animals:** The States may consider providing add-on coverage for crop loss due to attack by wild animals wherever the risk is perceived to be substantial and is identifiable.

General Exclusions: Losses arising out of war and nuclear risks, malicious damage and other preventable risks shall be excluded.

List of Insurance Companies

Department of Agriculture Cooperation & Farmers Welfare has designated/empanelled Agriculture Insurance Company of India (AIC) and some private insurance companies presently to participate in the Government sponsored agriculture /crop insurance schemes based on their financial strength, infrastructure, manpower and expertise etc. The empaneled insurance companies at present are:

- Agriculture Insurance Company of India Ltd
- ICICI-Lombard General Insurance Company Ltd
- HDFC -ERGO General Insurance Company Ltd.
- IFFCO-Tokio General Insurance Company Ltd.
- Cholamandalam MS General Insurance Company Ltd.
- Bajaj Allianz General Insurance Company Ltd.
- Reliance General Insurance Company Ltd.
- Future Generali India Insurance Company Ltd.
- Tata-AIG General Insurance Company Ltd.
- SBI General Insurance Company Ltd.
- Universal Sompo General Insurance Company Ltd.

The only government insurance company on this list, the Agriculture Insurance Company of India, exited the scheme after making large gains in the first year.

Unit of Insurance

The Scheme shall be implemented on an 'Area Approach basis' *i.e.,* **Defined Areas** for each notified crop for widespread calamities with the assumption that all the insured farmers, in a Unit of Insurance, to be defined as "Notified Area" for a crop, face similar risk exposures, incur to a large extent, identical cost of production per hectare, earn comparable farm income per hectare, and experience similar extent of crop loss due to the operation of an insured peril, in the notified area.

Defined Area (i.e., unit area of insurance) is Village/Village Panchayat level by whatsoever name these areas may be called for major crops and for other crops it may be a unit of size above the level of Village/Village Panchayat. In due course of time, the Unit of Insurance can be a pGeo-Fenced/Geo-mapped region having homogeneous Risk Profile for the notified crop.

For Risks of Localised calamities and Post-Harvest losses on account of defined peril, the Unit of Insurance for loss assessment shall be the affected insured field of the individual farmer.

Calendar of activity

Activity	Kharif	Rabi
Loaning period (loan sanctioned) for Loanee farmers covered on Compulsory basis.	April to July	October to December
Cut-off for receipt of Proposals of farmers (loanee & non-loanee)	31 July	31 December
Cut-off date for receipt of yield data	Within a month from final harvest	Within a month from final harvest

Revised operational guidelines for Pradhan Mantri Fasal Bima Yojna (PMFBY)

Government has modified operational guidelines for Pradhan Mantri Fasal Bima Yojna (PMFBY) which is being implemented from 1st October, 2018.

New provisions in the operational guidelines of PMFBY.

- Provision of Penalties/ Incentives for States, Insurance Companies (ICs) and Banks i.e. 12% interest rate to be paid by the Insurance Company to farmers for delay in settlement claims beyond two months of prescribed cut off date. Similarly, State Govt. have to pay 12% interest rate for delay in release of State share of Subsidy beyond three months of prescribed cut off date/submission of requisition by Insurance Companies.
- Detailed SOP for Performance evaluation of ICs and their de-empanelment.
- Inclusion of Perennial horticultural crops (on pilot basis) under the ambit of PMFBY (OGs of PMFBY envisages coverage of food and Oilseed crops and Annul Commercial & Horticultural crops).
- Inclusion of hailstorms in post harvest losses, besides unseasonal and cyclonic rainfalls.
- Inclusion of cloud burst and natural fire in localized calamities in addition to hailstorm, landslide, and inundation.
- Add on coverage for crop loss due to attack of wild animals on pilot basis with the additional financial liabilities of this provision to be borne by concerned state Govt.
- Mandatory capturing of Adhaar number — This would help in de-duplication.
- Target for Coverage to ICs especially of Non loanee farmers (10% incremental).
- Definition of Major Crops, Unseasonal rainfall and Inundation incorporated for clarity and proper coverage.
- Rationalization of premium release process: Release of Upfront premium subsidy based on 50% of 80% of total share of subsidy of corresponding season of previous year as

GOI/State subsidy at the beginning of the season- Companies need not provide any projections for the advance subsidy. Second Installment — balance premium based on approved business statistics on Portal for settlement of claims and final installment after reconciliation of entire coverage data on portal based on final business statistics on portal.

- States allowed to take decision for inclusion of crops having high premium for calculation of Ll calculation and for notification.
- Rationalization of methodology for calculation of TY.
- Moving average of best 5 out of 7 years for calculation of claim amount.
- Settlement of claims (Prevented sowing/ on account for Mid season adversity / Localized Claims) without waiting for Second installment of final subsidy.
- Yield based claims to be settled on the basis of subsidy provided on provisional business data.
- Separate Budget Allocation for Administrative expenses (atleast 2% of budget of scheme).
- Broad Activity wise seasonality discipline containing defined timelines for all major activities to streamlines the process of coverage, submission of yield data and early settlement of claims.
- District wise crop 'wise crop calendar (for major crops) to decide cut-off date for enrolment.
- Increased time for change of crop name for insurance - upto 2 days prior to cut-off date for enrolment instead of earlier provision of 1 month before cut-off date.
- More time to insured farmer to intimate individual claims — 72 hours (instead of 48 hours) through any stakeholders and directly on portal.
- Timeline for declaration of prevented sowing.
- Detailed SOP for dispute redressal regarding yield data/crop loss.
- Detailed SOP for claims estimation w.r.t. Add on products i.e. Mid season adversity, prevented/failed sowing, post harvest loss and localized claims.
- Detailed SOP for Area Correction factor.
- Detailed SOP for Multi picking crops.
- Detailed plan for publicity and awareness- earmarked expenditure-0.5% of Gross premium per company per season.
- Use of RST in clustering/Risk classification.
- Penalties/Incentives for States, ICs and Banks.
- Performance evaluation of ICs and their de-empanelment.

Pradhan Mantri Krishi Sinchai Yojna

Launched on 1st July, 2015 with the moto of "Har Khet Ko Paani", the Pradhan Mantri Krishi Sinchayee Yojna (PMKSY) is being implemented to expand cultivated area with assured irrigation, reduce wastage of water and improved water use efficiency.

The major objective of PMKSY is to achieve convergence of investments in irrigation at the field level, expand cultivable area under assured irrigation, improve on-farm water use efficiency to reduce wastage of water, enhance the adoption of precision-irrigation and other Water saving technologies (**More crop per drop**), enhance recharge of aquifers and introduce sustainable water conservation practices by exploring the feasibility of reusing treated municipal waste water for peri-urban agriculture and attract greater private investment in precision irrigation system.

PMKSY has been conceived amalgamating ongoing schemes *viz.* Accelerated Irrigation Benefit Programme (AIBP) of the Ministry of Water Resources, River Development & Ganga Rejuvenation (MoWR, RD&GR), Integrated Watershed Management Programme (IWMP) of Department of Land Resources (DoLR) and the On Farm Water Management (OFWM) of Department of Agriculture and Cooperation (DAC). The scheme will be implemented by Ministries of Agriculture, Water Resources and Rural Development. Ministry of Rural Development is to mainly undertake rain water conservation, construction of farm pond, water harvesting structures, small check dams and contour bunding etc. MoWR, RD&GR, is to undertake various measures. for creation of assured irrigation source, construction of diversion canals, field channels, water diversion/lift irrigation, including development of water distribution systems. Ministry of Agriculture will promote efficient water conveyance and precision water application devices like drips, sprinklers, pivots, rain-guns in the farm "(Jal Sinchan)", construction of micro-irrigation structures to supplement source creation activities, extension activities for promotion of scientific moisture conservation and agronomic measures.

Programme architecture of PMKSY will be, to adopt a decentralized State level planning and projectised execution' structure that will allow States to draw up their own irrigation development plans based on District Irrigation Plan (DIP) and State Irrigation Plan (SIP). It will be operative as convergence platform for all water sector activities including drinking water & sanitation, MGNREGA, application of science & technology etc. through comprehensive plan. State Level Sanctioning Committee (SLSC) chaired by the Chief Secretary of the State will be vested with the authority to oversee its implementation and sanction projects.

The programme will be supervised and monitored by an Inter-Ministerial National Steering Committee (NSC) will be constituted under the Chairmanship of Prime Minister with Union Ministers from concerned Ministries. A National Executive Committee (NEC) will be constituted under the Chairmanship of Vice Chairman, NITI Aayog to oversee programme implementation, allocation of resources, inter ministerial coordination, monitoring and performance assessment, addressing administrative issues etc.

PMKSY has the following programme components:

A. Accelerated Irrigation Benefit Programme (AIBP)

- To focus on faster completion of ongoing Major and Medium Irrigation including National Projects.

B. PMKSY (Har Khet ko Pani)

- Creation of new water sources through Minor Irrigation (both surface and ground water).
- Repair, restoration and renovation of water bodies; strengthening carrying capacity of traditional water sources, construction rain water harvesting structures (Jal Sanchay).
- Command area development, strengthening and creation of distribution network from source to the farm.
- Ground water development in the areas where it is abundant, so that sink is created to store runoff/ flood water during peak rainy season.
- Improvement in water management and distribution system for water bodies to take advantage of the available source which is not tapped to its fullest capacity (deriving benefits from low hanging fruits). At least 10% of the command area to be covered under micro/precision irrigation.
- Diversion of water from source of different location where it is plenty to nearby water scarce areas, lift irrigation from water bodies/rivers at lower elevation to supplement requirements beyond IWMP and MGNREGS irrespective of irrigation command.
- Creating and rejuvenating traditional water storage systems like Jal Mandir (Gujarat); Khatri, Kuhl (H.P.); Zabo (Nagaland); Eri, Ooranis (T.N.); Dongs (Assam); Katas, Bandhas (Odisha and M.P.) etc. at feasible locations.

C. PMKSY (Per Drop More Crop)

- Programme management, preparation of State/District Irrigation Plan, approval of annual action plan, Monitoring etc.
- Promoting efficient water conveyance and precision water application devices like drips, sprinklers, pivots, rain - guns in the farm (Jal Sinchan).
- Topping up of input cost particularly under civil construction beyond permissible limit (40%), under MGNREGS for activities like lining inlet, outlet, silt traps, distribution system etc.
- Construction of micro irrigation structures to supplement source creation activities including tube wells and dug wells (in areas where ground water is available and not under semi critical/critical /over exploited category of development) which are not supported under AIBP, PMKSY (Har Khet ko Pani), PMKSY (Watershed) and MGNREGS a s per block/district irrigation plan.
- Secondary storage structures at tail end of canal system to store water when available in abundance (rainy season) or from perennial sources like streams for use during dry periods through effective on - farm water management.
- Water lilting devices like diesel/electric/solar pumpsets including water carriage pipes, underground piping system.

- Extension activities for promotion of scientific moisture conservation and agronomic measures including cropping alignment to maximise use of available water including rainfall and minimise irrigation requirement (Jal sarankchan).
- Capacity building, training and awareness campaign including low cost publications, use of pico projectors and low cost films for encouraging potential use water source through technological, agronomic and management practices including community irrigation.
- The extension workers will be empowered to disseminate relevant technologies under PMKSY only after requisite training is provided to them especially in the area of promotion of scientific moisture conservation and agronomic measures, improved/ innovative distribution system like pipe and box outlet system, etc. Appropriate Domain Experts will act as Master Trainers.
- Information Communication Technology (ICT) interventions through NeGP - A to be made use in the field of water use efficiency, precision irrigation technologies, on farm water management, crop alignment etc. and also to do intensive monitoring of the Scheme.

D. PMKSY (Watershed Development)

- Effective management of runoff water and improved soil and moisture conservation activities such as ridge area treatment, drainage line 5 treatment, rain water harvesting, in-situ moisture conservation and other allied activities on watershed basis.
- Converging with MGNREGS for creation of water source to full potential in identified backward rainfed blocks including renovation of traditional water bodies.

■■■■

Chapter - 19 Management in Agricultural Extension Organisations

Agricultural extension activities are carried out by persons within an organization specifically designed and put together for extension activities. This will help to understand what an extension organization is. The concept of management and managerial skills required of a manager of an agricultural extension organisation.

Agricultural extension is the application of scientific research and knowledge to agricultural practices through farmer's education. Extension is also critical to move research from the lab to the field and to ensure a return on investment in research by translating new knowledge into innovative practices.

Concept of an Agricultural Extension Organization

Agricultural extension means different things to different people. To the farmers, it is a form of assistant to help improve their knowledge, efficiency, productivity, profitability and contribution to the good of their family, community and society. To the politicians, planners and policy makers, it is a policy instrument to increase agricultural production, to achieve national food security and at the same time help to alleviate rural poverty. To the practitioner, agricultural extension enhances and accelerates the spread of useful knowledge and technologies to rural people. These activities are expected to lead to increased and sustained productivity, increased income and well-being of farm people and to the promotion of national food security and economic growth. An agricultural extension organization is therefore a formalized intentional structure of roles or position targeted at achieving extension activities.

Concept of Management

Ever since people began forming groups to accomplish aims they could not achieve as individuals, managing has been essential to enter the coordination of individual efforts. Management is the process of designing and maintaining an environment in which individuals working together in groups efficiently accomplish selected aims. There are some necessary skills required of managers of extension organizations. We shall discuss three of them in this unit.

1. Managerial skills- Technical skills: Technical skill is knowledge and proficiencies in activities involving methods, process and procedures. Thus, it involves working with tools and specific techniques. Managers of extension organizations should thus be able to teach their staff specific techniques in doing their job.

2. Human skill : Human skill is the ability to work with people. It is cooperative effort, team works and the creation of an environment in which people feel secured and free to express their opinions. Managers of extension organizations should thus be competent in interpersonal relationships. They must be able to blend the efforts of subordinates who may differ in background, specialization and points of view into motivated teams of workers.

3. Conceptual and Design Skills : This is the ability of the managers to recognize significant elements in a situation and to understand the relationships among the elements. He must be able to findout the problem of the farmer in his social, economic and sometimes political context, he must be able to analyse the information at his disposal. The manager should be able to put forward practical solution to such problems. This is the aspect of design skill. The design skill is the ability of manager to solve problems.

It is necessary that formalized structures be set up to accomplish agricultural extension activities. Roles also need to be defined for people in such structures so that proper running and functioning of such organizations can be achieved. Such structures will therefore need people who have the necessary skills to coordinate such activities.

Principles of Management

Principles in management are fundamental truths. They are what are taught to be truths at a given time explaining relationships between two or more sets of variable, usually an independent variable and a dependent variable. Principle could be descriptive or predictive but not prescriptive. They describe how one variable relates to another, what all happen when these variables interact. They do not prescribe what people should do. The following are some of the commonly used principles:

1. Division of work : This is the specialization that economist consider necessary for efficiency in the use of labour. Division of work stems from the assumption that breaking down the work into smaller units of activities will help individuals to better focus on them as well as enable them to develop more skill in doing such activities.

2. Authority and responsibility : Authority is the power inherent in a position. It is a combination of official factors, deriving from the manager's position and personal factors. This is a mix of intelligence, experience, moral worth; past service etc. Responsibility arises from authority. It has to do with what he uses the authority to accomplish. Every manager must understand that he is not in position for any other reason but to ensure compliance to rules and regulations *vis-a-vis* accomplishment of organisational goals. He must use the power invested in that position maximally.

3. Discipline : Discipline is a respect for agreements, which are directed at achieving obedience, application, energy and the outward marks of respect. Discipline requires good superiors at all levels implying that those who will discipline must have a high moral justification for doing so. Discipline also requires that rules and regulations of the system are made explicit for all to know.

4. Unity of command : This means that employees should receive orders from one superior officer only. Thus, employees will be directly responsible to one superior officer per time. This will help prevent clash of instructions.

5. Unity of direction : This principle informs that each group of activities with the same objective must have one head and one plan. It relates to the organization of the 'body corporate' rather than to personnel. Though it does not mean all decision should be made at the top.

6. Subordination of individual to general interest : This refers to the submission of individual agenda and plan to the interests and welfare of the organization. Individuals must be ready to sacrifice their own ideas and interest to that of the organisation. This will enhance commitment and reduce interference of individual beliefs, perception etc with organisational goals or other workmates. Nevertheless, when the two are found to differ, management must reconcile them.

7. Remuneration : This principle suggests that remuneration and methods of payment should be fair and commensurate to services rendered. It should also affect the maximum possible satisfaction to employees and employer.

8. Centralization : This refers to the extent to which authority is concentrated or dispersed. Some agricultural extension organizations command more influence from management giving organisation a strong management style ready to face emergencies. Nevertheless, this style reduces and inhibits staff initiative and participation in management, which could also have detrimental consequences. However, individual circumstances will determine the degree that will give the best results.

9. Order : This is ensuring a place for everything and everyone. There must be materials and social order. People and resources must be organized/ well arranged to prevent chaos.

10. Equity managers : Equity managers should deal with subordinates in kindness and justice. This is to elicit loyalty and devotion from personnel. There must be fairness in treatment of staff based on established regulations to prevent favoritism

11. Stability of tenure : This principle is to ensure that staff are stable on their jobs and the rate of turnover is reduced to the barest minimum. High turnover poses a danger and increases costs and could also be both a cause and the effect of bad management.

12. Espirit de corps : This principle is that which believes —in union there is strength. It is an extension of the principle of unity of command. This principle emphasizes the need for teamwork and the importance of communication in obtaining it.

13. Initiative : This principle encouraged managers to permit their subordinates the opportunity to exercise their initiative. Initiative is conceived as the thinking out and execution of a plan. This is one of the keenest satisfactions for an intelligent man to experience. Subordinates should thus be believed in to be able to use their initiative for the good of the organization.

14. Scalar chain : This principle assumes a chain of superiors from the highest to the lowest ranks. This chain should not be departed from needlessly but could be short-circulated when to follow its compulsory would be detrimental.

Managerial functions in an agricultural extension organisation

Management of every organization has various statutory functions that they perform on a day-to-day basis or periodically such that the aims and objectives of the organization can be achieved. In agricultural extension organizations these functions do not differ from what obtains in other organization. In this unit, we shall consider those functions that are relavent to extension activities.

1. Planning : Planning involves the selection of mission and objectives and the action to achieve them. It requires decision making, that is, the choosing of future courses of action from among alternatives.

2. Organizing : Organizing involves establishing an intentional structure of roles for people to fill in an organization. It is intentional in the sense of making sure that all the tasks necessary to accomplish goals are assigned to people who can do them best. This organizational structure helps in creating an environment for human performance.

3. Staffing : Staffing involves filling and keeping filled, the position in the organizations structure. This is done by identifying work-force requirements; inventorying the people available and recruiting, selecting, placing, promoting, appraising, planning the careers of, compensating and training or otherwise developing both candidates and current job holders so that tasks are accomplished effectively and efficiently.

4. Leading or supervision : Leading or supervision involves influencing people so that they all contribute to organization and group goals. It has to do predominantly with the interpersonal aspect of managing. Leadership involves motivation, leadership styles, approaches and communication.

5. Controlling : Controlling is measuring and correcting individual and organizational performance to ensure that events conform to plans. It involves measuring performance against goals and plans, showing where deviations from standards exist and helping to correct them.

Theories Used in Agricultural Extension Organisations

Theories help us to make predictive judgments about what could happen in the occurrence of certain events. Among humans involved in work relationships, there is the need for those who manage them to be able to predict actions and reactions of their staff. This will enable better planning and decision-making. In this unit, we shall begin to examine theories used in the management of extension organization

A theory is an explicit and objective statement of which either by summary or speculation suggests that a relationship exist between data or knowledge and the reasoning from this observation. It is also a relationship between factors or the ordering of such facts in some meaningful way. Theory is a speculative idea or plan of how something might be done. It is the systematic statement of principles involved in some certain observed phenomenon.

Purpose of Theories

Theories are useful in agricultural extension organization management as they help managers and administrators in the execution of their duties. Theories of Extension organization serve the following purposes.

1. A basis for research.
2. A guideline to administration and organization of action.
3. It serves as a guide to the application of new knowledge.
4. A basis for decision making operations and action in organization.

5. It helps to understand, predict and adjust management behaviour more objectively and systematically.

However, no single theory can achieve the objectives set above. Thus, the different theories compliment the other in providing the framework for proper understanding of management operation. In this subunit, the six theories of management that will be examined are:

A. Functional theory

B. Decision making theory

C. Human relation theory

D. Fusion theory

E. Social process theory

F. Theory X, Y and Z

A. Functional theory

This theory is accredited to Fredrick Taylor. It explains administrative behaviour in terms of administration process and formal function of the administrator. The theory divides the function of an administrator into minute components of 1. Planning 2. Organizing 3. Staffing 4. Directing 5. Coordination 6. Reporting 7. Budgeting.

The functional theory assumes a division of labour among employers, which leads to three basic sub-concepts of :

1. **Division of labour** *i.e.* providing specific responsibilities to individuals *i.e.* making them responsible.
2. **Specialization** *i.e.* making individuals concentrates on one activity to acquire proficiency and expertise in it.
3. **Line-staff relationship** *i.e.* the relationship between the administrators (line) and the professional (staff).

Principles underlying the functional theory include

1. Specialization which helps to lead efficiency
2. Staff structure and hierarchy, which reveals each staff responsibility and line or authority. This also leads to efficiency.
3. Span of control, which helps to ascertain the number of persons that is economically profitable to the organization.
4. Adequate delegation of authority which helps a subordinate to be able to take responsibility in the absence of her superior.
5. Job evaluation by supervisors which helps management assess employee activities and thus remunerate accordingly.

B. Decision Making Theory

This theory established decision making in an organization as the complex of human association, events and activities leading to and including any conclusion for a programme of policy and/or operation. It helps to identify the best course of action from a set of objectives/ alternatives. It suggests that no single individual takes decision alone in any organization. The decision maker makes decisions within an environment. The goals or ends to be achieved must be stated. Relevant alternatives to accomplishing the goals must be stated which must be ordered in preference. The choice is finally made and the decision maker must be ready to accept responsibility for the decision taken.

C. Human Relations Theory

This theory suggests that interaction of people in terms of endeavour and the integration of people into work situation which motivates them to walk cooperatively and satisfactory. It evidence that individuals should be able to fit into another's shoe and be able to see things in another point 23 of view. Another aspect of the theory also suggest the primary roles of the administrator as being concerned with the causal forces which make a group or an individual think or being motivated to produce effectively and with satisfaction in his/her work. The theory believes that wages alone do not bring satisfaction to work. Those other nonphysical factors are divided into two (a) hygiene factors and (b) motivator factors.

D. Social Process Theory

This theory pioneered by Hawthrone is based on the philosophy that accepts organizational behaviours as a system of mutually interdependent variables. This theory sees an organization as a 25 complex of mutually dependent variables functioning and related to a social field. These variables include persons and their motives, attitudes, beliefs sentiments, jobs and activities, the physical work setting and the formal or informal organization itself. These factors and more others are related in complex whole. This theory makes organizations loose their distinctiveness and find meanings in terms of human behaviour and the operation of the total system. This theory thus emphasizes the interaction among the parts rather than the component parts in isolation of each other.

E. Fusion Theory

This theory postulated by Max Weber is based on the concepts of socialization and personalization. It believes that there is a simultaneous operation of the two concepts, where the organization attempts to restructure the individual as an agent for the accomplishment of its goals (socialization). On the other hands, the individual also tries to use the organization as a medium through which to achieve his personal objective that is, personalization. The administrator should therefore understand this and balance these two forces.

F. Theory X, Y and Z

This theory believes that scientific management and administration can predict and control human behaviour. The theories assert that the behaviour of the individual in any work environment depends on the type of administration style employed on the individual.

i. **Theory X** believer an average person hates work and should be threatened to perform at desired level. He avoids responsibility and prefers to be controlled, can only work if he is going to be punished or rewarded for doing so and in an industrial setting, less than his full potential is utilized.

ii. **Theory Y** is the opposite of theory X and believes an average person has positive attitude towards work, achieves the highest level of performance though self initiative, has a sense of responsibility, has basic needs and works through motivation, commitment, satisfaction and self direction. He actualized his dreams through self-initiative and personal development. Theory X is based on the principle of control and direction through the exercise of power and authority while theory Y is based on the creation of a situation in which the individual can achieve its goals through self-ambition and desire while also meeting organizational objectives.

iii. **Theory Z :** This believes there is no ideal situation where we have either situation of theory X and Y in operation. This theory reconciles the two theories. It believes the individual and organization are always in conflict because of the dual role of consumer and producer. The role of the administrator is thus to collate the aggregate demand of individual consumers and producers to meet that demand necessary for the attainment of organization goals. This theory assumes that by discipline of the organization and confidence of the individual in the organization, through a precise system of communication and the assurance that the organization is out to secure the workers welfare, the lapses of theory X and Y can be overcome.

Roles, Responsibilities and Attributes of different Agricultural Extension Personnel

Human resources form the bedrock of the agricultural extension system. This is because they are the ones that actually do the job of information and technology dissemination. It is thus necessary to clarify what each person does to enhance effectiveness and reduce clash of roles and responsibilities. This we shall look into in this unit.

A job description gives the details of what is expected from the current jobholder. On the other hand, key performance areas are specific and show the critical functions relevant at present and for the future to achieve the objectives. The identification of key performance areas helps in role clarity as well as in delegation of functions. This in turn aids in performance appraisal and training. Generally, four or five key areas for a job are identified. The core extension personnel of developing countries consist of village extension workers, subject-matter specialists and supervisory staff or extension officers. Examples of key performance areas of core extension personnel are given below.

Roles and Responsibilities of the Village Extension Workers

The following constitute the roles and responsibilities of the people in this category :

(1) make regular and systematic visits to villages and farms to develop rapport with the clientele and to understand their problems;

(2) undertake educational activities in the form of meetings, campaigns, demonstrations, field days, training sessions and exhibitions;

(3) provide advisory services to the farmers and solve their production problems.

Roles and Responsibilities of the Subject-Matter Specialists

The following constitute the roles and responsibilities of people in this category :

(1) Keep abreast of current recommendations and findings related to farm production by maintaining continuous contact with agricultural research stations;

(2) Provide feedback to the research system about farmers' problems which need solutions;

(3) Train and backstop village extension workers on the latest farm technology and help them in solving field problems.

Roles and Responsibilities of Supervisory Staff or Extension Officers

The following constitute the roles and responsibilities of people holding these positions:

(1) Plan, organize, coordinate and implement extension programmes and activities;

(2) Supervise and monitor the work of field staff, providing guidance, motivation, and evaluation of performance;

(3) Coordinate the programme with inter-and intradepartmental agencies.

Critical Attributes for Extension Personnel

The key performance areas indicate the important roles and contributions of different categories of extension personnel. Once the roles are delineated, they can be analysed to indicate the attributes, which can discriminate an effective from an ineffective role occupant. These critical attributes consist of qualities such as educational qualifications, skills, experience, physical characteristics, mental abilities, values and attitudes needed for extension. The critical attributes needed for field-level and supervisory extension staffs are necessary formal training in agriculture, practical skills and experience in farming, and knowledge of modern farm practices. Abilities in group dynamics, human relations and communication are also important. Basic skills related to management and leadership are needed by extension supervisors. Values and attitudes such as faith in rural people, commitment to agricultural development and concern for the whole community are important for all extension personnel.

■■■■

Chapter - 20 Organizational Structure of Extension Service

Organization and management play an important role in agriculture development. The organizational structure of an extension service plays a large role in what types of extension activities are carried out, how they are carried out and the recipient target groups and beneficiaries of extension.

There are two dominant approaches to the organization of extension services.

(I) identifies one particular institution devoted solely to extension, the second identifies an agricultural development organization with responsibilities for research, marketing and extension.

(II) The policy issue for consideration is whether to separate organizations dealing with different sub-sectors involved in production, such as crops, livestock, fruit, vegetables, fisheries and forestry and even youth development and family life.

It is known that the more specialized the individuals or groups, the higher the cost of effective communication and cooperation between them. However, agriculture is a highly technical multi-faceted field which often requires many kinds of specialized personnel in addition to generalists who comprehend whole farming systems.

The type of organizational structure, the principal factor which determines the success of any extension programme is the degree to which any specific project is adopted and supported by the target population. "Interventions which fail to sustain support and sponsorship within the national bureaucracy are destined to wither and dissipate. Production schemes which fail to win the confidence and enthusiasm of farmers will not generate food or revenue."

Presently, Indian agricultural extension has wide mandates and despite the pluralistic extension approaches, its coverage and use of service is limited; particularly in rainfed regions that are represented by marginal and smallholder farmers'. Hence, there is need to develop "need based" capacity building of small-scale men and women farmers, as well as. Gaining access to reliable information in increasing their productivity and profitability for livelihood improvements.

There are five major agricultural public sector extension systems devoted to extension work in India:

i. The Ministry of Agriculture at central level, including the Indian Council of Agricultural Research (ICAR) and the Directorate of Extension (DoE).

ii. State Department of Agriculture (DoA), as well as State Agricultural Universities (SAUs).

iii. The Department of Agriculture (DoA), Animal Husbandry (AH), Horticulture (DoH) and fisheries (DoF), as well as Krishi Vigyan Kendra (KVKs) and more recently the Agricultural Technology Management Agency (ATMA) at the district level.

iv. Also there is a wide variety of producers groups, including cooperatives and federations of milk, fruits, cotton, oilseeds, coconut, tobacco, spices etc.

v. Civil society organizations, such as the Non-governmental Organization (NGOs).

Roles of Extension in Different Institutions

The primary responsibility for agricultural extension in most countries lies with the government agency. The majority of countries for which information is available indicates that the ministry of Agricultuure is predominantly responsible for agriculture extension. The particular agency concerned varies greatly from country to country, and invariably generates a certain amount of controversy and competition. In some countries different agencies operate agriculture extension, and their services may or may not be coordinated.

In a typical agriculture extension agents tend to be trained in agronomy and may or may not have specific training in agriculture. Extension personnel are more likely to combine activities in agriculture with those in other types of agriculture. Farmers are therefore targeted for technology transfer of agriculture as well as technologies of agriculture. Thus there is greater potential for extension work with farming systems, such as integrated livestock-fish production and management of water resources for other agricultural activities.

In cases where responsibility for agriculture extension is within an agency for natural resources, there is a likelihood that the aquaculture extension agents are also used for functions such as licensing, regulatory work and enforcement. Thus their overall emphasis is often on preserving natural resources, and not commercial production or exploitation for home consumption.

Universities in India are involved in agriculture extension. University extension services may or may not coordinate their activities with government agencies. In cases where they do, universities play an extremely important role in training extension agents, "link workers", and producers. Universities may also have research activities in agriculture and allied subject which generate technologies which can be transferred to farmers directly and quickly.

For the design of the extension programme a survey was carried out to identify any social factors which would affect the project. The survey identified factors such as other enterprises on the farm, resources available for agriculture, levels of previous experience or interest and economic and financial limitations. The project evaluation indicated the need to have specific and well-defined goals for project participants. The survey indicated that the produce for home consumption was most appropriate, but this could possibly lead to subsequent commercialization.

Military institutions in many countries can play roles in extension. In general extension provided by the military is not formally linked to other national extension activities, but often national extension agents request specific help from military institutions in the form of the loan of heavy equipment.

Private voluntary agencies and organizations are playing an increasingly important role in agriculture extension in many countries. Their typical approach is small-scale and grass-roots oriented, with significant inputs from local producers at various stages of projects. In general, "promotores" are recruited and trained to work directly with a certain number of farmers.

Agencies of foreign governments, which provide bilateral technical assistance, also provide useful agriculture extension services in many of their projects. These foreign government agencies generally train their own personnel who then work through a host country institution with host-country counterparts in extension activities. For example, United States of America has the Agency for International Development (USAID) and Peace Corps, and other countries, such as Norway (Norwegian Agency for International Development, or NORAD), Sweden (Swedish International Development Authority, or SIDA), France, Federal Republic of Germany (Deutsche Gesellschaft fur Technische Zusammenarbeit GmbH, or GTZ), Italy, Japan (Japan International Cooperation Agency, or JICA), United Kingdom (Overseas Development Administration, or ODA), among others, send technical advisors and volunteers who work in some type of extension-related activities. Invariably these individuals have high levels of education and/or important practical and technical skills useful in many countries. Problems arise, however, from language and cultural barriers, and sometimes through introducing technologies and institutional or organizational structures not well adapted to host-country conditions.

Foreign assistance agencies may also have different priorities and criteria for designing and implementing projects. The national goals for a particular project and those of the bilateral agency may be in conflict, and the policies of an agency may differ widely depending on the policies of its respective government in power. Thus, individual projects which overlap different political administrations may be designed under the first with certain goals and objectives but then evaluated under a second administration, with vastly different criteria, policies, and objectives.

Multilateral organizations, such as the Food and Agriculture Organization of the United Nations (FAO), the Inter-American Development Bank (IADB), and the Asian Development Bank (ADB), among others, include agriculture extension in many of their projects. FAO, for example, through the financial support of the United Nations Development Programme (UNDP) and bilateral agencies, has established a number of regional and national training centres for agriculture technicians.

Linkages between Extension and Research Efforts

Many countries, however, do not have resources to engage in research. In those countries which do, institutional and professional linkages may not exist to provide effective communication between researchers and extensionists. Often inter-institutional rivalry, competition for funding, and communication difficulties impede their development. Effective technology transfer requires firm linkages between research and extension. Technology is developed through research, and extensionists must have access to researchers to know and understand the latest technology to ensure its successful transmittal to farmers.

Research personnel often view themselves as being above extensionists, as they have more formal education and postgraduate degrees. Usually they are willing to provide information to extension personnel, but rarely willing to receive information back from extension personnel and adjust the problems they study accordingly. This causes a breakdown in communication and coordination.

The classic picture of extension is one where the researcher communicates his research information to the extensionist for transmittal to producers. One factor, however, is often overlooked. The researcher invariably has a professional stake in the quality of the research, and the extension worker has a personal one. Consequently an individual farmer, who has built up a personal relationship with the agent, may try a new practice based simply on faith in that particular agent. If the research was of poor quality, or erroneous, then the farmer pays the price, and the extension agent receives the blame. The researcher is removed from these personal ties. Many researchers, in fact, do not have sufficient regard for the implications of their research, or hasty publication and pronouncement of results. Consequently it is important for extension workers to know researchers and trust their abilities, and for researchers to understand that farmers need practical solutions and research in their problems.

The survey indicated that there were no institutional vehicles in most countries for linking research and extension activities. Conventions, conferences, meetings with researchers and extension agents, and task forces of researchers and extension agents seem to be institutionalized and developed. Other respondents indicated few linkages between research and extension activities.

Evaluation for Effective Management

The definition of evaluation is and how useful it is to assess how effective the chosen administrative, organisational and supervisory procedures. Evaluations are often criticised for not having a sufficient evidence base. This is often because monitoring data is impossible to collect during the evaluation itself. Many evaluations are designed based on the assumption that a reasonable quantity and quality of data is available.

Meaning and Importance of Evaluation

Evaluation involves checking the effectiveness of the tools and methods used in an extension programme to bring about change in a community or organisation. It gives a basis for adjusting and improving the extension programme. This process helps the extension agents determine the degree to which they are accomplishing what they set out to do and gives them the opportunity to make a report to the public about their activities or the value of a programme.

Models of extension programme evaluation

There are different models for extension programme evaluation. Some of the major models are:

1. Expert model : This model depends on experts making judgements on the strengths and limitations of a programme based on interviews, document analysis, their own perspectives or standards that are set by stakeholders.

2. Goal-free model : The main focus of this model is for external evaluators to identify environmental and farming conditions and then to compare the needs identified by the extension programme to what people experience as a result of extension programme implementation. This model is used to identify the needs that the programme has not met and then uses this information to make changes.

3. Attainment of objectives model : This model assumes that the programme's success depends on achieving the outcomes compared to the original goals and objectives. In this evaluation model, it is important to outline measurable objectives from the beginning and then gather information that will indicate how much progress has been made in meeting these objectives.

4. Management decision model : This model allows managers to gather the information they need to make decisions. Stakeholder participation is very important in this model because their decisions are influenced by evaluation. One disadvantage of this model is that major stakeholders hold the most decision making power and their decisions are given preference over the views of less powerful stakeholders.

5. Naturalistic model : This model is valuable in making positive collaborative changes when conflict resolution skills are combined with evaluation. This is because this model assumes that programmes are negotiations between stakeholders and the outcome of evaluation is based on disagreements about objectives, expectations, problems, opportunities, policies, procedures, and suggested changes in methods or activities.

6. Experimental model : This approach is used to determine if the changes to programme outcomes were caused by programme inputs or from other influences. Evaluating the success of a programme depends on comparing the outcomes of an experimental group (e.g. farmers who received training) to the outcomes of a control group (e.g. farmers who did not receive training). There are some ethical implications with using this model, because it is not ethical to withhold services from people who could benefit from them.

7. Participatory evaluation model : This model relies on the extensionists and farmers identifying an ongoing major situation that needs to be changed, then to critically reflect on how their own activities might be maintaining the negative situation. After new assumptions and practices are introduced, results are validated or invalidated based on whether a positive change happened when the new approach was adopted.

Designing Extension Evaluation

Steps in Evaluation

There are following steps in the evaluation process:

1. Choose an extension project or activity;
2. Find out more about the background of the target clients;
3. Analyse the extension teaching methods;
4. Drawing up an action plan;
5. Determine sample sizes and collect data;
6. Arrange and edit the collected data;
7. Interpret the facts;
8. Plan how to use the findings to improve the programme or make decisions.

Types of Evaluation

1. Evaluation of extension in the value chain: In the value chain, adding value to products or services means that people will be prepared to pay more for a product or service that they think has benefitted them than one that they felt did not benefit them. Evaluation in the value chain assesses whether the value added is the best value for money or if there are areas that need to be improved to maximise the value of a product or service. Value for money should reflect results that are valued by farmers as users of extension services.

2. Evaluation accountability in extension : An element of evaluation is reporting the progress of activities of a programme to stakeholders or the public. Receiving feedback or complaints in response to these progress reports ensures that the programme managers will be accountable to the stakeholders. It also adds information that allows programme managers to make changes to improve the programme.

3. Evaluation of human resources in extension : Employee performance appraisals are a type of evaluation used in human resource management to assess if an employee is adding value to the organisation, if they are entitled to a promotion or increased remuneration for their efforts or if they are underperforming. 109 Part of the New Extensionist Learning Kit Evaluation of extension sustainability and ownership. In donor-funded interventions, sustainability can only be proven to be successful or not sometime after external funding has been discontinued.

Challenges of India's Extension System

In current scenario, where a numbers of stakeholders are involving in agricultural extension, hence, opportunity to reach a greater number of farmers is increasing. In this context, private sector is incorporating extension services within existing service provisions and experimenting with ICT. But inherent challenges each sector faces in reaching different farmers means that partnership and coordination between sectors will best serve the interests of farmers. Hence, addressing of current challenges is necessary.

Public Sector Extension System

- In Indian extension system, information flow within public sector moves linearly, with content focusing on transfer of technology for enhancing agricultural production. A wider definition of agricultural extension, beyond improving crop productivity, has not been embraced. Information flow is supply-driven and not need based or area specific, therefore farmers see the quality of information provided by public extension staff as a major shortcoming.
- There are also insufficient funds for operational costs, training, and capacity development, which limits the activities and continual development of the extension staff. However, it was experienced that there are about 90,000 on the job, which is an adequate number of extension workers for the number of farmers (about 130 million).
- Various line departments at the state and district levels have been criticized for working in isolation, with weak linkages and rare partnerships. The research–extension link has been criticized for not absorbing or using feedback from farmers and extension staff. Extension personnel and farmers are passive actors, and scientists have limited exposure to field realities.
- Numerous components of public-sector extension system suffer from duplication of programs, without convergence. While ATMA is pushed as the platform through which

the multiple agencies can converge, the implementation difficulties are proving great for effective integration, with shortages of both personnel and funds.

Private Sector Extension System

To diffuse agricultural information directly to farmers, private-sector examples are developing context-specific models and using ICT tools.

- In India, private sector is playing an important task in extension services. The public sector recognizes this, with the policy framework for agricultural extension referring to the need for public extension services not to crowd out private services. Additionally, policy framework for agricultural extension notes that "public extension by itself cannot meet specific needs of various regions and different classes of farmers".
- In the pluralistic extension systems, private sector can provide services related to proprietary goods, while the public sector can provide extension services related to public goods, which tend not to be addressed by private-sector firms.
- Furthermore, private sector serves a corporate interest, working with individual farmers, so social capital is not built. Moreover, private extension can only work well if farmers are willing and able to pay indirectly through the sale of inputs.
- The private sector could serve the needs of medium size and commercial farmers, while the public sector could work in remote areas, which are currently not serviced well. This sort of system would require Public Private Partnerships (PPP) that currently does not exist in India. It would mean changes in the way the public sector views and interacts with the private sector. Relying on the public sector may also be difficult for remote and resource-poor farmers, considering the existing problems and poor reach of the public sector in those areas.

Civil Society (NGO) Extension Systems

Within information value chains, the capacity of farmers to articulate their needs will influence their ability to obtain information they need. Considering a large number of marginal and small land holdings in India, both the FIGs and SHGs can play important roles in articulating the needs of men and women farmers to knowledge intermediaries. These FIGs/SHGs can operate side by side with either NGOs or the public sector. However, challenges exist in both sectors.

- Public capacity to build FIGs and SHGs is limited, while NGOs, which are not numerous, rely on donor funds and would need public support to develop the technical skills to facilitate groups.
- Building social capital is critical in overall agricultural development strategies for reducing rural poverty. In a large country like India, through public extension system, meeting of scientists with farmers and visit of farmers to research institutes is a time consuming and difficult task.
- Both FIGs/SHGs are already emerging as an effective mechanism for both the transfer of technologies and the empowerment of the rural poor. Adoption of this approach can reduce the extension cost and workload of extension functionaries.

- For that, ICTs could be useful tools to increase connectivity between the various FIGs/SHGs and different extension approaches. Covering the whole country where diversities and complexities are prevalent in agriculture as well as mentally makeup for converting into social capital (especially of the downtrodden, like landless laborers, smallholders, rural women etc.) is a herculean task.
- For harnessing the ICTs, there are many challenges prevalent like insufficient infrastructure, sustainability aspect, capacity development etc., which needs to be addressed.
- Capacity building of SHGs/FIGs and promoting development of leadership and management skills are utmost needed so that farmers can demand information they need. It is therefore an important component of agricultural extension approaches.

■■■■

Chapter - 21 Role of Women in Agriculture

Women play a significant and crucial role in agricultural development and allied fields. The nature and extent of women's involvement in agriculture varies greatly from place to place and region to region. Women have played and continue to play a key role in the conservation of basic life support systems such as land, water, flora and fauna. They have protected the health of the soil through organic recycling and promoted crop security through the maintenance of varietal diversity and genetic resistance.

Rural women perform numerous labour intensive jobs such as weeding, hoeing, grass cutting, picking, cotton stick collection, separation of seeds from fibre, keeping of livestock and its other associated activities like milking, milk processing, preparation of ghee, etc. Rural women often manage complex households and pursue multiple livelihood strategies. Their activities typically include producing agricultural crops, tending animals, processing and preparing food, working for wages in agricultural or other rural enterprises, collecting fuel and water, engaging in trade and marketing, caring for family members and maintaining their homes.

Agriculture is the backbone of the Indian economy. Women play a vital role in building this economy. Over the years, there is a gradual realization of the key role of women in agricultural development and their vital contribution in the field of agriculture, food security, horticulture, processing, nutrition, sericulture, fisheries, and other allied sectors.

Women first domesticated crop plants and thereby initiated the art and science of farming. While men went out hunting in search of food, women started gathering seeds from the native flora and began cultivating those of interest from the point of view of food, feed, fodder, fibre and fuel. Women have played and continue to play a key role in the conservation of basic life support systems such as land, water, flora and fauna. They have protected the health of the soil through organic recycling and promoted crop security through the maintenance of varietal diversity and genetic resistance.

Poultry farming is one of the major sources of rural economy. The rate of women in poultry farming at household level is the central in poultry industry. Even though rural women are not using modern management techniques, such as vaccination and improved feed, but their poultry enterprise is impressive. Every year, income from poultry farming has been rising. In order to generate more and more income, rural women often sell all eggs and poultry meat and left nothing for personal use. Due to poverty and lack of required level of proteins most of women have got a very poor health. Most of women suffer from malnutrition.

Multi-dimensional role of women

Agricultural activities: Sowing, transplanting, weeding, irrigation, fertilizer application, plant protection, harvesting, winnowing, storing etc.

Domestic activities: Cooking, child rearing, water collection, fuel wood gathering, household maintenance etc.

Allied activities: Cattle management, fodder collection, milking, pultry raring, fish farming etc.

Despite women's extensive and varied participation in agriculture, they continue to have less access than do men to modern agricultural inputs. As a result, their farm work is labor intensive and yields meager economic returns. The bulk of the development projects for rural women over the 1976-85 period provided training in traditional female skills (e.g. sewing, cooking, and crafts), credit programs for microenterprises, and income-generation schemes to bring rural women into the market economy.

Agricultural extension efforts should help women improve food production while allowing them to shift more of their labor to export production. Similarly, changes in legal, financial, and educational systems must be undertaken in order to enhance women's social and economic contributions to rural development in the long term. There is a need to examine carefully the implications of land tenure laws and regulations for women. Educational policies and funding must be changed to reflect the very high social and economic returns to women's primary education and literacy.

There are following contribution of women in agriculture:

- Women contribute to two-thirds of the world's work hours, produce 50 per cent of the world's food supplies.
- Women work in fields, take care of families and manage household.
- Despite the services rendered by women in the family and work place, they make up for nearly 70 per cent of the world's poor and more than 65 per cent of the illiterates.

1. Women do in Agriculture and Rural Employment

Women make important contributions to the agricultural and rural economies of all regions of the world. However, the exact contribution both in terms of magnitude and of its nature is often difficult to assess and shows a high degree of variation across countries and regions.

2. Women in the Agricultural Labour Force

The female share in India has remained steady at just over 30 percent. These very large countries mask changes in some smaller countries where the female share of the agricultural labour force appears to have increased dramatically, now exceeding 50 percent in Bangladesh.

3. Time Spent in Agricultural Activities

Women are typically involved in majority of labour for agricultural work such as land preparation, fertilizer application, weeding, harvesting, threshing and storage. However, women overwhelmingly provide the greatest proportion of household time spent on food processing and preparation.

4. Women's Contribution to Agricultural Production

Women play a significant role in the agricultural labour force and in agricultural activities, although to a varying degree. Consequently their contribution to agricultural output is undoubtedly extremely significant, although difficult to quantify with any accuracy. It has often been claimed that women produce 60-80 per cent of food. However, assigning contributions to agricultural outputs by gender is problematic because in most agricultural households both men and women are involved in crop production.

5. Women as Livestock Keepers

Within pastoralist and mixed farming systems, livestock play an important role in supporting women and in improving their financial situation, and women are heavily engaged in the sector. They share responsibility with men and children for the care of animals, and particular species and types of activity are more associated with women than men. For example, women often have a prominent role in managing poultry and dairy animals and in caring for other animals that are housed and fed within the homestead. When tasks are divided, men are more likely to be involved in constructing housing and herding of grazing animals, and in marketing of products. The influence of women is strong in the use of eggs, milk and poultry meat for home consumption and they often have control over marketing and the income from these products. In some countries small-scale pig production is also dominated by women. Female-headed households are as successful as male-headed households in generating income from their animals, although they tend to own smaller numbers of animals. Ownership of livestock is particularly attractive to women in societies where access to land is restricted to men, while the role of women in small-scale livestock production is well recognized.

6. Women in Fisheries and Aquaculture

A large number of women are directly engaged, full-time or part-time, in the fishery primary sector. In addition to this majority of women are employed in the secondary sector, including post-harvest activities. Women may comprise up to 30 percent of the total employment in fisheries, including primary and secondary activities. Women are more commonly occupied in subsistence and commercial fishing from small boats and canoes in coastal or inland waters. Women also contribute as entrepreneurs and provide labour before, during and after the catch in both artisanal and commercial fisheries. For example, in West Africa, the so called "Fish Mamas" play a major role. They usually own capital and are directly and vigorously involved in the coordination of the fisheries chain, from production to sale of fish. In India, women represented a share of 24 percent, of all fishers and fish farmers.

The most significant role played by women in both artisanal and industrial fisheries is at the processing and marketing stages. In some countries, women have become important entrepreneurs in fish processing; in fact, most fish processing is performed by women, either in their own household-level industries or as wage labourers in the large-scale processing industry.

7. Women and Unpaid Household Responsibilities

Women are responsible for most of the household and child-rearing activities as well rearing of small livestock, although norms differ by culture and over time. This additional work burden is unpaid and limits women's capacity to engage in income-earning activities, which often require a minimum fixed time before being profitable. Furthermore, the nature of tasks, such as caring for children and elderly household members, requires women to stay near the home, thus limiting options to work for a wage.

Women are also usually responsible for child care and household chores. Depending on the household structure and size, these tasks may be extremely time intensive. Time-allocation studies have shown that women work significantly more than men.

Women from low-income groups and living in areas with limited facilities, spend long hours on water and fuel collection, food preparation and other domestic and child care activities to compensate for poor infrastructure. Women (and daughters) are responsible for about 65 percent of all transport activities in rural households, including travel for firewood, water and transport to the grinding mill. Deforestation leads women to collect firewood from increasingly further distances from the homestead .

The multi – dimensional work as performed by farm women are further discussed below.

Agriculture

Mainly rural women are engaged in agricultural activities in three different ways depending on the socio-economic status of their family and regional factors. They work as:

- Paid labourers.
- Cultivator doing labour on their own land.
- Managers of certain aspects of agricultural production by way of labour supervision and the participation in post harvest operations.

The types of agricultural activities taken up by women include the following :

- Sowing
- Nursery management
- Transplanting
- Weeding
- Irrigation
- Fertilizer application
- Plant protection
- Harvesting, threshing, winnowing, storing etc.

Livestock

Livestock is the primary livelihood activity used to meet household food needs as well as supplement farm incomes. It is a common practice in the rural areas to give an animal as part of a women's dowry. Studies have revealed rural women earn extra income from the sale of milk and animals. Mostly women are engaged in cattle management activities such as

- Cleaning of animal and sheds

- Watering of cattle
- Milking the animals
- Fodder collection
- Preparing dung cakes
- Collection of farm yard manure

Except grazing, all other livestock management activities are predominantly performed by women. Men, however, share the responsibility of taking care of sick animals. It is evident that the women are playing a dominant role in the livestock production and management activities.

Poultry

Poultry farming is one of the major sources of rural economy. The rate of women participation in poultry farming at household level is central in poultry industry.

Fish/aqua culture

Farm women plays an important role for production to sale of fishes and aqua products. They are also actively engaged for processing of fishes.

Feminisation of Agriculture

Economic Survey 2017-18 says that with growing rural to urban migration by men, there is 'feminisation' of agriculture sector, with increasing number of women in multiple roles as cultivators, entrepreneurs, and labourers. Globally, there is empirical evidence that women have a decisive role in ensuring food security and preserving local agro-biodiversity. Rural women are responsible for the integrated management and use of diverse natural resources to meet the daily household needs. This requires that women farmers should have enhanced access to resources like land, water, credit, technology and training which warrants critical analysis in the context of India. In addition, the entitlements of women farmers will be the key to improve agriculture productivity. The differential access of women to resources like land, credit, water, seeds and markets needs to be addressed.

With women predominant at all levels-production, pre-harvest, post-harvest processing, packaging, marketing of the agricultural value chain, to increase productivity in agriculture, it is imperative to adopt gender specific interventions. An 'inclusive transformative agricultural policy' should aim at gender-specific intervention to raise productivity of small farm holdings, integrate women as active agents in rural transformation, and engage men and women in extension services with gender expertise.

Significance of Women in Indian Agriculture

Background Society is the base for overall development of a nation. Society's structure depends entirely upon the status and condition of its women. The condition of women was far better in ancient India. In early period Aryans treated men and women alike in all religious rituals and social functions. But condition of women has declined gradually. Gender is the word used especially for the female population of the society. Many historians believe that it was woman who first domesticated crop plants and thereby initiated the art and science of farming. In those primitive days, when men went out hunting in search of food, women started gathering seeds from the native flora and began cultivating. Since then, women have played a key role in crop production, livestock production, horticulture, post harvest operations, agro/social forestry, fisheries, etc. Significance Women's contribution to agriculture, whether it is in subsistence farming or commercial agriculture, when measured in number of tasks performed and time spent, is greater than men. Their contribution in agriculture is aptly highlighted by a micro study conducted in Indian Himalayas which found that on a one hectare farm, a pair of bullocks work for 1060 hours, a man for 1212 hours and a woman for 3485 hours. Generally, operations performed by men are those that entail use of machinery and animals. Contrary to this, women always rely on manual labour using only their own energy. In rural India, the percentage of women who depend on agriculture for their livelihood is as high as 84%. Women make up about 33% of cultivators and about 47% of agricultural labourers. These statistics do not account for work in livestock, fisheries and various other ancillary forms of food production in the country. In 2009, 94% of the female agricultural labour force in crop cultivation were in cereal production, while 1.4% worked in vegetable production, and 3.72% were engaged in fruits, nuts, beverages, and spice crops. Not only are women overworked, their work is more arduous than that undertaken by men. Further, since women's work is largely based on human energy it is considered unskilled and hence less productive. On this basis, women are invariably paid less wage despite their working harder and for longer hours. This makes the woman's work invisible. In India, rural women are extensively involved in agricultural activities and form the most important productive workforce. They are engaged in agricultural activities in three different ways depending on the socio-economic status of their family and regional factors. They work as: Paid Labourers, Cultivator doing labour on their own land and Managers of certain aspects of agricultural production by way of labour supervision and the participation in post harvest operations. Rural India is witnessing a process which could be described as "feminization of agriculture".

Role of Cooperatives for Women Farmers in India

Cooperatives have been long seen as a social institution providing partnership, solidarity and resources to women farmers as well as tackle gender inequality. In India they have had quite a success. In many instances in which women are barred from participation, women only cooperatives are critical in empowering and educating. Yet female participation in cooperatives is still relatively low and some argue because men are still seen as primarily in charge of agriculture and income generation. Only 7.5% of women participate in cooperatives as compared to 92.5% of men. Of India's 450,000 cooperatives with a membership of 204.5 million, there are only 8,171 women cooperatives with a total membership of 693,000 women. Despite that, women-only cooperatives, which include cooperative banks, stores, food vendors, have done quite well and provided a whole range of services to their members. In India, with a view to involve women in the process of decision-making in local self-governing bodies including cooperatives, a 33% representation has been instituted and in a number of states all boards of directors have women serving on them. International organizations such as the Self Employed Women's Association (SEWA) have been working quite successfully in India with partners to form a membership of 1.24 million women in India. Fifty four percent of members are agricultural workers.

Government Initiatives for Women Farmers in Indian Agriculture

Efforts have been made by department of agriculture to incorporate gender issues into development agenda and ensure full and equitable participation of women in all agricultural development programmes. In an effort to 'engender agricultural process' the DOAC implemented special programmes/ schemes for women in agriculture sector, where a composite package of training, extension, input, managerial and entrepreneurial support was provided to selected farm women beneficiaries. Different approaches were followed in all these projects to provide a complete support package to women farmers. The technical back-stopping was provided through a blend of 'skilled-based' capacity building exercises coupled with extension support through "On field Demonstration", "Mahila-Goshties" and "Study Tours".

Problems of Women in Agriculture Sector in India

1. Gender Division of Labour in India

In India, the typical work of the female agricultural labour or cultivator is limited to less skilled jobs, such as sowing, transplanting, weeding and harvesting, that often fit well within the framework of domestic life and child-rearing. Many women also participate in agricultural work as unpaid subsistence labour. According to United Nations Human Development Report only 32.8% of Indian women formally participate in the labour force, a rate that has remained steady since 2009 statistics. By comparison men constitute 81.1%.

2. Literacy

An estimated 52-75% of Indian women engaged in agriculture are illiterate, an education barrier that prevents women from participating in more skilled labour sectors. In all activities there is an average gender wage disparity, with women earning only 70 per cent of men's wage. Additionally, many women participate in agricultural work as unpaid subsistence labour. The lack of employment mobility and education render the majority of women in India vulnerable, as dependents on the growth and stability of the agricultural market.

3. Time Allocation

In addition to rigorous agricultural work that is undervalued and underpaid, women are also responsible for the well-being of the household. They care for their children, provide nutrition or usually take part in subsistence agriculture, and do chores around the house. Based on time allocation, which pinpoint exactly how a woman's hours are spent throughout the week, Indian women spend about 25 hours in a week doing household chores and five hours in caring and community work. Besides the 30 hours of unpaid work, women spend the same amount of time as men carrying out agricultural work. Daughters typically supplement or substitute for mother's unpaid work around the household. Considered female tasks, the opportunity cost of girls' time for school is higher than that of sons. Girls do significantly more housework than boys, which compromises their schooling.

4. Property Ownership Rights of Indian Women

Critical resources such as land are also unevenly distributed by gender. Women seldom enjoy property ownership rights directly in their names. They have little control over decisions made in reference to land. Even with land in their names, they may not have actual decision-making power in terms of cropping patterns, sale, mortgage and the purchase of land. In India only 14.9% of households are female headed. Access to credit is difficult, since women lack many of the prerequisites for lending such as assets or ownership of property. Without access to capital or household decision making abilities women lack the resources that are necessary for their labour stability and stability of their households. Land ownership opportunities also have a critical impact on human development with freedom from violence. According to a 2005 study of marital violence and property ownership, 49% of property less women experience physical violence and 84% experienced psychological abuse. Ownership rights saw a drastic decrease in violence. Among women who owned both land and house there was only 7% physical violence and 16% psychological abuse.

5. Rural Credit

Moreover, the access of women to the most crucial input "credit" is limited since they are not land owners, the credit flow generally goes in the name of male members *i.e.* owners. Credit policies of various banks and other organizations need orientation towards women by granting them the status of a producer.

How to Improved the Status of Farm Women

Following are the key points, which will help in creating respectable role of women in Indian agriculture

1. Skill empowerment by training in the area of various operations

i. Field operations
ii. Conservation of biodiversity
iii. Nutritional bio-security
iv. Vocational training
v. Organic farming

2. Technology development for women

a. Designing of tools for various field operations
b. Animal husbandry
    i. Artificial insemination.
    ii. Veterinary knowledge
    iii. Care of animals against disease infection
c. For side income
    i. Mushroom cultivation
    ii. Floriculture
    iii. Kitchen gardening

3. Creation of self help groups

a. For financial support
b. For generation of employment

4. Projection of contribution of women by collecting and analyzing data

a. Collection and display of data
b. Projection of successful women in agriculture
c. Representation of their contribution in economic terms

5. Providing financial powers

Giving representation in land holdings

Challenges Faced by Female Farmers in Agriculture

1. Women perform all un-mechanized agricultural tasks and perform multiple tasks which add more burden to them due to lack of equipment and appropriate modern technology.
2. Women have limited access to use of productive resources.
3. Women have unequal land rights. They hardly enjoy land ownership rights directly in their names. Limited rights or access to arable land further limits livelihood options and exacerbates financial strain on women, especially in women- headed households.
4. Women have little control over decision making process, either inside home or outside home. Without access to capital or household decision making abilities women lack the resources that are for their labour stability and stability of their household.
5. Few women holding of agricultural productive resources such as land, animals and machinery.
6. Poor women farmers are less able to purchase technology to adapt to climate change due to lack of access to credit and agricultural services. They often have low productivity due to an inability to invest in things such as improved seeds and soil replenishment.
7. Women farmers in agricultural sector suffer from high illiteracy rate among them. They do not know their legal rights.
8. Women earn less wages, especially in joint, informal and private sector.
9. Miss applying some laws and regulations in favour of women such heritage legislation.
10. Lack of market intelligence and inadequate information put women farmers under unfavourable situation with weak bargaining power with the buyers.

Suggestions for Improving Status of farm Women

These are the following suggestions for the recognition of women contribution in agriculture:

1. Priority must be given to women in accessing credit on soft terms from banks and other financial institutions for setting up their business, for buying properties, and for house building.

2. Recognition of labour work of working women in the rural economy may be accounted in monetary terms.
3. More facilities should be provided to poor rural women for land, agricultural and livestock extension services.
4. Measures should be taken to enhance women's literacy rates. A separate education policy for women may serve the purpose.
5. Women must be involved in decision-making bodies that have the potential to introduce structural changes. This action will bring some changes in the gender relations in the society.
6. Minimum and equal wages should be fixed by state government for women agricultural labours and the rates should be reviewed periodically.
7. Conscious efforts are needed for training of female agricultural workers in the rural areas in alternative skills.
8. Women must be aware regarding their existing rights, access to judicial relief and redress, removing discrimination through legal reforms, and providing legal aid, assistance and counselling.
9. There is a need to make available cheap credit to needy agricultural women to start self employment and as a result of that our rural women may get additional opportunity for gainful employment.
10. Addition to above, some alternatives employment programmes and opportunities should be provided to agricultural women labour in rural areas.

■■■■